SCIENTIFIC AMERICAN EXPLORES BIG IDEAS

The Math Behind Science, Society, and Technology

The Editors of *Scientific American*

SCIENTIFIC AMERICAN EDUCATIONAL PUBLISHING

New York

Published in 2023 by Scientific American Educational Publishing
in association with **The Rosen Publishing Group**
2544 Clinton Street, Buffalo NY 14224

Contains material from Scientific American®, a division of Springer Nature America, Inc., reprinted by permission, as well as original material from The Rosen Publishing Group®.

First Edition

Scientific American
Lisa Pallatroni: Project Editor

Rosen Publishing
David Kuchta: Compiling Editor
Michael Moy: Senior Graphic Designer

Cataloging-in-Publication Data
Names: Scientific American.
Title: The Math behind science, society, and technology / Scientific American.
Description: New York : Scientific American Educational, 2023. | Series: Scientific american explores big ideas | Includes bibliographic information, glossary and index.
Identifiers: ISBN 9781684169368 (pbk) | ISBN 9781684169375 (library bound) | ISBN 9781684169382 (ebook)
Subjects: LCSH: Mathematics– Juvenile literature | Physics– Juvenile literature | Sociology– Juvenile literature
Classification: LCC HM529 S35 2023 | DDC 300 –dc23

Manufactured in the United States of America
Websites listed were live at the time of publication.

Cover: agsandrew/Shutterstock.com

CPSIA Compliance Information: Batch # SACW23.
For Further Information contact Rosen Publishing at 1-800-237-9932.

CONTENTS

INTRODUCTION

Galileo Galilei once observed that "the great book of the universe" is written in mathematical language, "without whose help it is impossible to comprehend a single word of it; without which one wanders in vain through a dark labyrinth." Mathematics lies behind everything, from the cryptic calculations of the quantum physicist to the averages and percentages of the ardent sports fan. To truly understand the world, Galileo knew, you need math. The articles in this book reveal the world in numbers—not the dry numbers of multiplication tables or algebraic equations, but the living numbers that keep our bodies growing, our economy running, and our Earth spinning.

Section 1, "Problems and Solutions," shows you that not everything is settled in mathematics, that even math—the fundamental language of the entire universe—has puzzles still to be solved. The articles in Section 2, "Society," reveal the math behind where we travel within urban areas, how we vote, and what divides us. Section 3, "Technology," examines how the high-tech industry uses math to solve mysteries and keep passwords secret. Section 4, "Health," shows us how small calculations can have large impacts on our and our planet's health. Section 5, "Education," looks at the pros and cons of how math is taught in schools. And Section 6, "Culture," shows the ways math is all around us, from the games we can play with it to the meters beneath our feet.

From the moment your alarm goes off in the morning, through the calories you eat during the day, and the sheep you count at night, our lives are measured by numbers. Whether you're a math geek or not, the articles in this book will entertain as much as they enlighten. For without math, we might still be wandering in vain through a dark labyrinth, as Galileo warned—because without math, we might still think that the Sun moves around the Earth.

Section 1: Problems and Solutions

Hilbert Walked so the Clay Mathematics Institute Could Run

By Evelyn Lamb

I've been reading up on Hilbert's problems recently. They're the 23 problems presented by German mathematician David Hilbert at the International Congress of Mathematicians held in Paris in 1900. (To be more precise, Hilbert presented 10 of them in a talk at the congress. The full list was published later. Mathematicians don't make a distinction between the 10 presented in person and the rest.)

Hilbert was one of the giants of mathematics at the turn of the twentieth century, and his assessment of the most tantalizing open problems in mathematics has influenced the course of mathematics research in the time since.

I started looking into the Hilbert problems because of a project I was working on that touched on the Millennium Prize Problems. This is a list of seven problems put forth in 2000 by the Clay Mathematics Institute as seven of the most important unsolved problems in mathematics at the start of the new millennium. I had thought of the millennium prize problems as a sort of update of Hilbert's list, but I was less familiar with his questions, so I wanted to learn more.

Each Millennium Prize Problem has a $1 million prize attached. Solve any one, and you could be a millionaire. (Or, like the only person to solve one of them so far, you could reject the prize and leave the research world entirely.*) Probably because of the lucrative purse attached to each problem, the Clay Institute has detailed guidelines for how prizes are decided and awarded and official write-ups of each problem by experts in the field. The status of each problem is well-defined. (The Poincaré conjecture is solved, and the other six are unsolved.)

Compared to the Millennium Prize Problems, Hilbert's list looks a bit like a term paper that got thrown together at the last minute by a student who forgot about the deadline. Some problems are too easy: the third problem had actually been solved before Hilbert stated it, though the solution was not published at the time; Max Dehn, who actually published the solution, solved it very quickly after Hilbert stated it. Some problems are ill-posed: the fourth problem is sometimes described as "too vague to have a solution," and the 23rd problem is just "Further development of the methods of the calculus of variations." Sorting the problems into "solved" and "unsolved" lists is a bit of a mess. The 10th problem has clearly been solved. The eighth problem, which includes the Riemann hypothesis, is clearly still open. But mathematicians have varying perspectives about the extent to which, for example, the second problem should be considered solved. In my research for the article, I consulted several books and websites, all of which had slightly different points of view on which questions were solved and which were still open. I could not help but think that in the most charitable interpretation, the amateurish Hilbert problems had paved the way for the smooth, professional Millennium Prize Problems. (Hence the title of this post.)

But maybe that description is unfair to Hilbert. Why should a list of unsolved problems in mathematics be tidy? After all, the problems are unsolved, which means we don't know whether or how they will be solved or if their solutions will show us that we were asking the wrong questions in the first place. Hilbert himself wrote, "It is difficult and often impossible to judge the value of a problem correctly in advance; for the final award depends upon the gain which science obtains from the problem." From another perspective, the Millennium Problems are corporate and sanitized compared to Hilbert's more organic, sprawling list.

The money clouds things as well. An open-ended question like "further development in the calculus of variations" is not reasonable when there's a lucrative prize for coming to some defined endpoint, but the calculus of variations has certainly developed a lot since 1900, in part because of Hilbert's challenge.

After I got past seeing Hilbert's list as sloppy and nebulous, I began to appreciate the way these problems—either in spite or because they are a little messy—fostered several fruitful lines of inquiry in 20th century mathematics. Sometimes the most important part was that they inspired mathematicians to figure out what the "right" questions were in the first place. Which combinations of axioms, sets of numbers or shapes, and quantifiers lead to the most interesting questions to look at? Even resolved problems, like the 10th, have opened up areas of research that continue to enthrall mathematicians.

Nothing says the Millennium Problems are not having the same effect. Only one of them has been solved, but partial results for others continue to come in. Will this smaller collection of more tightly-focused problems inspire as much innovation as Hilbert's did? There's no way to know now.

In any human endeavor, necessity is the mother of invention. In math, problems are often the source of that necessity. Hilbert said it as well:

> The deep significance of certain problems for the advance of mathematical science in general and the important role which they play in the work of the individual investigator are not to be denied. As long as a branch of science offers an abundance of problems, so long is it alive; a lack of problems foreshadows extinction or the cessation of independent development. Just as every human undertaking pursues certain objects, so also mathematical research requires its problems. It is by the solution of problems that the investigator tests the temper of [their] steel; [they find] new methods and new outlooks, and [gain] a wider and freer horizon.

*Grigori Perelman, who proved the Poincaré conjecture, refused both the Clay Institute prize money and a Fields Medal. The money was used to establish the Poincaré Chair, which funds 6- and 12-month visiting positions at the Institut Henri Poincaré in Paris for promising researchers. My spouse received one of these positions a few years ago, so my assessment should be considered very biased, but funding several short-term positions seems like a better way to encourage math research than giving a big pile of money to one person.

About the Author

Evelyn Lamb is a freelance math and science writer based in Salt Lake City, Utah.

Is the Mathematical World Real?

By Kelsey Houston-Edwards

When I tell someone I am a mathematician, one of the most curious common reactions is: "I really liked math class because everything was either right or wrong. There is no ambiguity or doubt." I always stutter in response. Math does not have a reputation for being everyone's favorite subject, and I hesitate to temper anyone's enthusiasm. But math is full of uncertainties–it just hides them well.

Of course, I understand the point. If your teacher asks whether 7 is a prime number, the answer is definitively "yes." By definition, a prime number is a whole number greater than 1 that is only divisible by itself and 1, such as 2, 3, 5, 7, 11, 13, and so on. Any math teacher, anywhere in the world, anytime in the past several thousand years, will mark you correct for stating that 7 is prime and incorrect for stating that 7 is not prime. Few other disciplines can achieve such incredible consensus. But if you ask 100 mathematicians what explains the truth of a mathematical statement, you will get 100 different answers. The number 7 might really exist as an abstract object, with primality being a feature of that object. Or it could be part of an elaborate game that mathematicians devised. In other words, mathematicians agree to a remarkable degree on whether a statement is true or false, but they cannot agree on what exactly the statement is about.

One aspect of the controversy is the simple philosophical question: Was mathematics discovered by humans, or did we invent it? Perhaps 7 is an actual object, existing independently of us, and mathematicians are discovering facts about it. Or it might be a figment of our imaginations whose definition and properties are flexible. The act of doing mathematics actually encourages a kind of dual philosophical perspective, where math is treated as both invented and discovered.

This all seems to me a bit like improv theater. Mathematicians invent a setting with a handful of characters, or objects, as well as a few rules of interaction, and watch how the plot unfolds. The actors rapidly develop surprising personalities and relationships, entirely independent of the ones mathematicians intended. Regardless of who directs the play, however, the denouement is always the same. Even in a chaotic system, where the endings can vary wildly, the same initial conditions will always lead to the same end point. It is this inevitability that gives the discipline of math such notable cohesion. Hidden in the wings are difficult questions about the fundamental nature of mathematical objects and the acquisition of mathematical knowledge.

Invention

How do we know whether a mathematical statement is correct or not? In contrast to scientists, who usually try to infer the basic principles of nature from observations, mathematicians start with a collection of objects and rules and then rigorously demonstrate their consequences. The result of this deductive process is called a proof, which often builds from simpler facts to a more complex fact. At first glance, proofs seem to be key to the incredible consensus among mathematicians.

But proofs confer only conditional truth, with the truth of the conclusion depending on the truth of the assumptions. This is the problem with the common idea that consensus among mathematicians results from the proof-based structure of arguments. Proofs have core assumptions on which everything else hinges—and many of the philosophically fraught questions about mathematical truth and reality are actually about this starting point. Which raises the question: Where do these foundational objects and ideas come from?

Often the imperative is usefulness. We need numbers, for example, so that we can count (heads of cattle, say) and geometric objects such as rectangles to measure, for example, the areas of fields.

Sometimes the reason is aesthetic–how interesting or appealing is the story that results? Altering the initial assumptions will sometimes unlock expansive structures and theories, while precluding others. For example, we could invent a new system of arithmetic where, by fiat, a negative number times a negative number is negative (easing the frustrated explanations of math teachers), but then many of the other, intuitive and desirable properties of the number line would disappear. Mathematicians judge foundational objects (such as negative numbers) and their properties (such as the result of multiplying them together) within the context of a larger, consistent mathematical landscape. Before proving a new theorem, therefore, a mathematician needs to watch the play unfold. Only then can the theorist know what to prove: the inevitable, unvarying conclusion. This gives the process of doing mathematics three stages: invention, discovery and proof.

The characters in the play are almost always constructed out of simpler objects. For example, a circle is defined as all points equidistant from a central point. So its definition relies on the definition of a point, which is a simpler type of object, and the distance between two points, which is a property of those simpler objects. Similarly, multiplication is repeated addition, and exponentiation is repeated multiplication of a number by itself. In consequence, the properties of exponentiation are inherited from the properties of multiplication. Conversely, we can learn about complicated mathematical objects by studying the simpler objects they are defined in terms of. This has led some mathematicians and philosophers to envision math as an inverted pyramid, with many complicated objects and ideas deduced from a narrow base of simple concepts.

In the late 19th and early 20th centuries a group of mathematicians and philosophers began to wonder what holds up this heavy pyramid of mathematics. They worried feverishly that math has no foundations–that nothing was grounding the truth of facts like 1+1=2. (An obsessive set of characters, several of them struggled with mental illness.) After 50 years of turmoil, the expansive project failed to produce a single, unifying answer that satisfied all the original goals, but it spawned various new branches of mathematics and philosophy.

Some mathematicians hoped to solve the foundational crisis by producing a relatively simple collection of axioms from which all mathematical truths can be derived. The 1930s work of mathematician Kurt Gödel, however, is often interpreted as demonstrating that such a reduction to axioms is impossible. First, Gödel showed that any reasonable candidate system of axioms will be incomplete: mathematical statements exist that the system can neither prove nor disprove. But the most devastating blow came in Gödel's second theorem about the incompleteness of mathematics. Any foundational system of axioms should be consistent—meaning, free of statements that can be both proved and disproved. (Math would be much less satisfying if we could prove that 7 is prime and 7 is not prime.) Moreover, the system should be able to prove—to mathematically guarantee—its own consistency. Gödel's second theorem states that this is impossible.

The quest to find the foundations of mathematics did lead to the incredible discovery of a system of basic axioms, known as Zermelo-Fraenkel set theory, from which one can derive most of the interesting and relevant mathematics. Based on sets, or collections of objects, these axioms are not the idealized foundation that some historical mathematicians and philosophers had hoped for, but they are remarkably simple and do undergird the bulk of mathematics.

Throughout the 20th century mathematicians debated whether Zermelo-Fraenkel set theory should be augmented with an additional rule, known as the axiom of choice: If you have infinitely many sets of objects, then you can form a new set by choosing one object from each set. Think of a row of buckets, each containing a collection of balls, and one empty bucket. From each bucket in the row, you can choose one ball and place it in the empty bucket. The axiom of choice would allow you to do this with an infinite row of buckets. Not only does it have intuitive appeal, it is necessary to prove several useful and desirable mathematical statements. But it also implies some strange things, such as the Banach-Tarski paradox, which states that you can break a solid ball into five pieces and reassemble those pieces into two new solid balls, each equal in size to the first. In other words, you can double the ball. Foundational assumptions are judged by the structures they produce, and the axiom of choice implies many important statements but also brings extra baggage. Without the axiom of choice, math seems to be missing crucial facts, though with it, math includes some strange and potentially undesirable statements.

The bulk of modern mathematics uses a standard set of definitions and conventions that have taken shape over time. For example, mathematicians used to regard 1 as a prime number but no longer do. They still argue, however, whether 0 should be considered a natural number (sometimes called the counting numbers, natural numbers are defined as 0,1,2,3... or 1,2,3..., depending on who you ask). Which characters, or inventions, become part of the mathematical canon usually depends on how intriguing the resulting play is—observing which can take years. In this sense, mathematical knowledge is cumulative. Old theories can be neglected, but they are rarely invalidated, as they often are in the natural sciences. Instead mathematicians simply choose to turn their attention to a new set of starting assumptions and explore the theory that unfolds.

Discovery

As noted earlier, mathematicians often define objects and axioms with a particular application in mind. Over and over again, however, these objects surprise them during the second stage of the mathematical process: discovery. Prime numbers, for example, are the building blocks of multiplication, the smallest multiplicative units. A number is prime if it cannot be written as the product of two smaller numbers, and all the nonprime (composite) numbers can be constructed by multiplying a unique set of primes together.

In 1742 mathematician Christian Goldbach hypothesized that every even number greater than 2 is the sum of two primes. If you pick any even number, the so-called Goldbach conjecture predicts that you can find two prime numbers that add up to that even number. If you pick 8, those two primes are 3 and 5; pick 42, and that is 13 + 29. The Goldbach conjecture is surprising because although primes were designed to be multiplied together, it suggests amazing, accidental relations between even numbers and the sums of primes.

An abundance of evidence supports Goldbach's conjecture. In the 300 years since his original observation, computers have confirmed that it holds for all even numbers smaller than 4×10^{18}. But this evidence is not enough for mathematicians to declare Goldbach's conjecture correct. No matter how many even numbers a computer checks, there could be a counterexample—an even number that is not the sum of two primes—lurking around the corner.

Imagine that the computer is printing its results. Each time it finds two primes that add up to a specific even number, the computer prints that even number. By now it is a very long list of numbers, which you can present to a friend as a compelling reason to believe the Goldbach conjecture. But your clever friend is always able to think of an even number that is not on the list and asks how you know that the Goldbach conjecture is true for that number. It is impossible for all (infinitely many) even numbers to show up on the list. Only a mathematical proof—a logical argument from basic principles demonstrating that Goldbach's conjecture is true for every even number—is enough to elevate the conjecture to a theorem or fact. To this day, no one has been able to provide such a proof.

The Goldbach conjecture illustrates a crucial distinction between the discovery stage of mathematics and the proof stage. During the discovery phase, one seeks overwhelming evidence of a mathematical fact—and in empirical science, that is often the end goal. But mathematical facts require a proof.

Patterns and evidence help mathematicians sort through mathematical findings and decide what to prove, but they can also be deceptive. For example, let us build a sequence of numbers: 121, 1211, 12111, 121111, 1211111, and so on. And let us make a conjecture: all the numbers in the sequence are not prime. It is easy to gather evidence for this conjecture. You can see that 121 is not prime, because $121 = 11 \times 11$. Similarly, 1211, 12111 and 121111 are all not prime. The pattern holds for a while–long enough that you would likely get bored checking–but then it suddenly fails. The 136th element in this sequence (that is, the number 12111...111, where 136 "1"s follow the "2") is prime.

It is tempting to think that modern computers can help with this problem by allowing you to test the conjecture on more numbers in the sequence. But there are examples of mathematical patterns that hold true for the first 10^{42} elements of a sequence and then fail. Even with all the computational power in the world, you would never be able to test that many numbers.

Even so, the discovery stage of the mathematical process is extremely important. It reveals hidden connections such as the Goldbach conjecture. Often two entirely distinct branches of math are intensively studied in isolation before a profound relation between them is uncovered. A relatively simple example is Euler's identity, $e^{i\pi} + 1 = 0$, which connects the geometric constant π with the number i, defined algebraically as the square root of -1, via the number e, the base of natural logarithms. These surprising discoveries are part of the beauty and curiosity of math. They seem to point at a deep underlying structure that mathematicians are only beginning to understand.

In this sense, math feels both invented and discovered. The objects of study are precisely defined, but they take on a life of their own, revealing unexpected complexity. The process of mathematics therefore seems to require that mathematical objects be simultaneously viewed as real and invented–as objects with concrete, discoverable properties and as easily manipulable inventions of mind. As philosopher Penelope Maddy writes, however, the duality makes no difference to how mathematicians work, "as long as double-think is acceptable."

Real or Unreal?

Mathematical realism is the philosophical position that seems to hold during the discovery stage: the objects of mathematical study—from circles and prime numbers to matrices and manifolds—are real and exist independently of human minds. Like an astronomer exploring a far-off planet or a paleontologist studying dinosaurs, mathematicians are gathering insights into real entities. To prove that Goldbach's conjecture is true, for example, is to show that the even numbers and the prime numbers are related in a particular way through addition, just like a paleontologist might show that one type of dinosaur descended from another by showing that their anatomical structures are related.

Realism in its various manifestations, such as Platonism (inspired by the Greek philosopher's theory of Platonic forms), makes easy sense of mathematics' universalism and usefulness. A mathematical object has a property, such as 7 being a prime number, in the same way that a dinosaur might have had the property of being able to fly. And a mathematical theorem, such as the fact that the sum of two even numbers is even, is true because even numbers really exist and stand in a particular relation to each other. This explains why people across temporal, geographical and cultural differences generally agree about mathematical facts—they are all referencing the same fixed objects.

But there are some important objections to realism. If mathematical objects really exist, their properties are certainly very peculiar. For one, they are causally inert, meaning they cannot be the cause of anything, so you cannot literally interact with them. This is a problem because we seem to gain knowledge of an object through its impact. Dinosaurs decomposed into bones that paleontologists can see and touch, and a planet can pass in front of a star, blocking its light from our view. But a circle is an abstract object, independent of space and time. The fact that π is the ratio of the circumference to the diameter of a circle is not about a soda can or a doughnut; it refers to an abstract mathematical circle, where distances are exact and the points on the circle are infinitesimally small. Such a perfect circle is causally inert and seemingly inaccessible. So how can we learn facts about it without some type of special sixth sense?

That is the difficulty with realism–it fails to explain how we know facts about abstract mathematical objects. All of which might cause a mathematician to recoil from his or her typically realist stance and latch onto the first step of the mathematical process: invention. By framing mathematics as a purely formal mental exercise or a complete fiction, antirealism easily skirts problems of epistemology.

Formalism, a type of antirealism, is a philosophical position that asserts that mathematics is like a game, and mathematicians are just playing out the rules of the game. Stating that 7 is a prime number is like stating that a knight is the only chess piece that can move in an L shape. Another philosophical position, fictionalism, claims that mathematical objects are fictions. Stating that 7 is a prime number is then like stating that unicorns are white. Mathematics makes sense within its fictional universe but has no real meaning outside of it.

There is an inevitable trade-off. If math is simply made up, how can it be such a necessary part of science? From quantum mechanics to models of ecology, mathematics is an expansive and precise scientific tool. Scientists do not expect particles to move according to chess rules or the crack in a dinner plate to mimic Hansel and Gretel's path–the burden of scientific description is placed exclusively on mathematics, which distinguishes it from other games or fictions.

In the end, these questions do not affect the practice of mathematics. Mathematicians are free to choose their own interpretations of their profession. In *The Mathematical Experience*, Philip Davis and Reuben Hersh famously wrote that "the typical working mathematician is a Platonist on weekdays and a formalist on Sundays." By funneling all disagreements through a precise process—which embraces both invention and discovery—mathematicians are incredibly effective at producing disciplinary consensus.

About the Author

Kelsey Houston-Edwards is a mathematician and journalist. She formerly wrote and hosted the online show PBS Infinite Series.

The Top Unsolved Questions in Mathematics Remain Mostly Mysterious

By Rachel Crowell

In the year 2000, mathematicians released a list of the top seven unsolved problems in the field. Answering them would offer major new insights in fundamental mathematics and might even have real-world consequences for technologies such as cryptography.

But big questions in math have not often attracted the same level of outside interest that mysteries in other scientific areas have. When it comes to understanding what math research looks like or what the point of it is, many folks are still stumped, says Wei Ho, a mathematician at the University of Michigan. Although people often misunderstand the nature of her work, Ho says it does not have to be difficult to explain. "My cocktail party spiel is always about elliptic curves," she adds. Ho often asks partygoers, "You know middle school parabolas and circles? Once you start making a cubic equation, things get really hard.... There are so many open questions about them."

One famous open problem called the Birch and Swinnerton-Dyer conjecture concerns the nature of solutions to equations of elliptic curves, and it is one of the seven Millennium Prize Problems that were selected by the founding scientific advisory board of the Clay Mathematics Institute (CMI) as what the institute describes as "some of the most difficult problems with which mathematicians were grappling at the turn of the second millennium." At a special event held in Paris on May 24, 2000, the institute announced a prize of $1 million for each solution or counterexample that would effectively resolve one of these problems for the first time. Rules revised in 2018 stipulate that the result must achieve "general acceptance in the global mathematics community."

The 2000 proclamation gave $7 million worth of reasons for people to work on the seven problems: the Riemann hypothesis, the

Birch and Swinnerton-Dyer conjecture, the P versus NP problem, the Yang-Mills existence and mass gap problem, the Poincaré conjecture, the Navier-Stokes existence and smoothness problem, and the Hodge conjecture. Yet despite the fanfare and monetary incentive, after 21 years, only the Poincaré conjecture has been solved.

An Unexpected Solution

In 2002 and 2003 Grigori Perelman, a Russian mathematician then at the St. Petersburg Department of the Steklov Mathematical Institute of the Russian Academy of Sciences, shared work connected to his solution of the Poincaré conjecture online. In 2010 CMI announced that Perelman had proved the conjecture and, along the way, had also solved the late mathematician William Thurston's related geometrization conjecture. (Perelman, who rarely engages with the public, famously turned down the prize money.)

According to CMI, the Poincaré conjecture focuses on a topological question about whether spheres with three-dimensional surfaces are "essentially characterized" by a property called "simple connectivity." That property means that if you encase the surface of the sphere with a rubber band, you can compress that band—without tearing it or removing it from the surface—until it is just a single point. A two-dimensional sphere or doughnut hole is simply connected, but a doughnut (or another shape with a hole in it) is not.

Martin Bridson, a mathematician at the University of Oxford and president of CMI, describes Perelman's proof as "one of the great events of, certainly, the last 20 years" and "a crowning achievement of many strands of thought and our understanding of what three dimensional spaces are like." And the discovery could lead to even more insights in the future. "The proof required new tools, which are themselves giving far-reaching applications in mathematics and physics," says Ken Ono, a mathematician at the University of Virginia.

Ono has been focused on another Millennium Problem: the Riemann hypothesis, which involves prime numbers and their distribution. In 2019 he and his colleagues published a paper in the *Proceedings of the National Academy of Sciences USA* that reexamined an old, formerly abandoned approach for working toward a solution. In an accompanying commentary, Enrico Bombieri, a mathematician at the Institute for Advanced Study in Princeton, N.J., and a 1974 winner of mathematics' highest honor, the Fields Medal, described the research as a "major breakthrough." Yet Ono says it would be unfounded to describe his work as "anything that suggests that we're about to prove the Riemann hypothesis." Others have also chipped away at this problem over the years. For instance, mathematician "Terry Tao wrote a nice paper a couple years ago on [mathematician Charles] Newman's program for the Riemann hypothesis," Ono says.

Progress on What Won't Work

The fact that just one of the listed problems has been solved so far is not surprising to the experts–the puzzles are, after all, long-standing and staggeringly difficult. "The number of problems that have been solved is one more than I would expect" to see by now, says Manjul Bhargava, a mathematician at Princeton University and a 2014 Fields medalist. Bhargava himself has reported multiple recent results connected to the Birch and Swinnerton-Dyer conjecture, including one in which he says he and his colleagues "prove that more than 66 percent of elliptic curves satisfy the Birch and Swinnerton-Dyer conjecture."

None of the problems will be easy to solve, but some may prove especially intractable. The P versus NP problem appears so difficult to solve that Scott Aaronson, a theoretical computer scientist at the University of Texas at Austin, calls it "a marker of our ignorance." This problem concerns the issue of whether questions that are easy to verify (a class of queries called NP) also have solutions that are easy to find (a class called P). Aaronson has written extensively about the P versus NP problem. In a paper published in 2009 he and Avi Wigderson, a mathematician and computer scientist at the Institute for Advanced Study and one of the winners of the 2021 Abel Prize, showed a new barrier to proving that the P class is not the same as the NP class. The barrier that Aaronson and Wigderson found is the third one discovered so far.

"There's a lot of progress on showing what approaches will not work," says Virginia Vassilevska Williams, a theoretical computer scientist and mathematician at the Massachusetts Institute of Technology. "Proving that P [is] not equal to NP would be an important stepping-stone toward showing that cryptography is well founded," she adds. "Right now cryptography is based on unproved assumptions," one of which is the idea that P is not equal to NP. "In order to show that you cannot break the cryptographic protocols that people need in modern computers," including ones that keep our financial and other online personal information secure, "you need to at least prove that P is not equal to NP," Vassilevska Williams notes. "When people have tried to pin me down to a number," Aaronson says, "I'll give a 97 percent or 98 percent chance that P is not equal to NP."

Climbing Mount Everest

Searching for solutions to the prize problems is similar to trying to climb Mount Everest for the first time, Ono says. "There are various steps along the way that represent progress," he adds. "The real question is: Can you make it to base camp? And if you can, you still know you're very far."

For problems such as the Birch and Swinnerton-Dyer conjecture and the Riemann hypothesis, Ono says, "surely we're at Nepal"—one of the countries of departure for climbing the mountain—"but have we made it to base camp?" Mathematicians might still need additional "gear" to trek to the peak. "We're now trying to figure out what the mathematical analogues are for the high-tech tools, the bottles of oxygen, that will be required to help us get to the top," Ono says. Who knows how many obstacles could be sitting between current research and possible solutions to these problems? "Maybe there are 20. Maybe we're closer than we think," Ono says.

Despite the difficulty of the problems, mathematicians are optimistic about the long term. "I hope very much that while I'm president of the Clay institute, one of them will be solved," says Bridson, who notes that CMI is in the process of strategizing about how to best continue raising awareness about the problems. "But one has to accept that they're profoundly difficult problems that may continue to shape mathematics for the rest of my life without being solved."

About the Author

Rachel Crowell is a Midwest-based writer covering science and mathematics.

Gnarly, Centuries-Old Mathematical Quandaries Get New Solutions

By Rachel Crowell

A series of unsolved puzzles in number theory called Diophantine problems date back to 3,700 years ago. Over the years mathematicians have whittled away at them, and recent work has made significant progress on some–and showed others to be just as uncrackable as ever.

Researchers have been using tools from geometry to tackle the problems, which are named after Diophantus, a third-century Greek mathematician. They involve determining which solutions exist for polynomial equations such as $x^n + y^n = z^n$. Mathematicians aim to find out if there are any integer or rational solutions to the equations. For instance, for $x^2 + y^2 = z^2$, infinitely many such solutions exist.

Diophantine geometry is the field of math focused on the relationship between the number theory properties of an equation, such as its rational or integer solutions, and "the geometric properties, like the topology of the set of complex solutions to the equation," says David Corwin, a mathematician at Ben Gurion University of the Negev in Israel.

It is surprising "how little we know about Diophantine geometry compared to other fields in mathematics," says Bjorn Poonen, a mathematician at the Massachusetts Institute of Technology. For instance, he notes that although mathematicians know that the number 20 can be written as the sum of three cubes, as in $3^3 + 1^3 + (-2)^3 = 20$, whether the number 114 can be expressed as the sum of three cubes remains an open problem.

The "Dark Side"

For some Diophantine problems, mathematicians' focus might seem obsessively narrow. Why expend substantial effort to determine whether 114 can be written as the sum of three cubes? Kiran Kedlaya, a mathematician at the University of California, San Diego, says that for many Diophantine puzzles that are simple to state, "the problem itself is not so central ... but the techniques that are needed to solve it are very central."

This property is not uncommon in math. The famous quandary known as Fermat's Last Theorem, for instance, is also more important because of the techniques developed to solve it than for the problem itself, Kedlaya says, "which doesn't have much in the way of direct consequences for number theory." The tools used to attack it, however, include key advances in algebraic number theory in the late 19th century, as well as in modular forms in the early 20th century. "Those [developments] are incredibly important for solving lots of problems in modern number theory," he says, including questions connected to cryptography.

"The very simplest problems tend to be the motivation that lead us to develop techniques that we can then use to solve the problems that really tell us a lot," he says. For example, the Serre Uniformity Problem, which relates to Kedlaya's research, concerns a special type of mathematical curve called a modular curve. However, "the consequences of it are quite deep and the techniques that we're using to apply it" to different cases are rooted in earlier work on the Fermat problem, Kedlaya notes.

Still, some of the Diophantine problems are more intractable than others. "Many researchers in the field try to develop new methods for solving Diophantine equations," Poonen says, "but I work also on the 'dark side' by trying to prove that some classes of problems are unsolvable."

New Tools for Old Problems

Rather than employ tools from geometry and other fields to solve specific Diophantine problems, it might have been possible to develop computer programs to solve the general case of such problems instead. But mathematicians Martin Davis, Yuri Matiyasevich, Hilary

Putnam and Julia Robinson, showed that finding full solutions to these problems isn't as simple as tasking a computer with searching for them. Their work culminated in a 1970 theorem that answered German mathematician David Hilbert's famous 10th problem. That problem focused on finding an algorithm for determining whether, for some system of polynomial equations with integer coefficients, there exists a solution in the integers, Kedlaya notes. In thinking that such an algorithm could be found, "Hilbert was an optimist," Kedlaya says. "Hilbert was big on trying to take care of general classes of problems."

But the Matiyasevich theorem, which is also called the DPRM theorem or the MRDP theorem, showed that such an algorithm doesn't exist. The discovery means that the "general problem of that type is, of course, intractable," and individual instances of these problems can be "very hard to solve," Kedlaya says.

Curiously, Corwin notes that for polynomial equations (or systems of such equations) in multiple variables, no one knows whether an algorithm can be found for determining whether rational solutions exist. "It's anyone's guess," he says. Poonen has worked to show that such a general method for finding solutions in rational numbers is impossible.

For some of these ancient questions, including ones posed by Diophantus himself, "we're only just now developing methods that can help answer them," says Jennifer Balakrishnan, a mathematician at Boston University. For example, a problem from Diophantus's book *Arithmetica* concerns whether positive, rational solutions for x and y exist such that the equation $y^2 = x^8 + x^4 + x^2$ is satisfied. Though Diophantus provided a solution, which is $x = ½$ and $y = 9/16$, Balakrishnan says that until 1998, it was unknown how many other solutions existed. In a doctoral thesis at the University of California, Berkeley, Joseph Wetherell presented techniques for answering this question.

More recently, Balakrishnan and her collaborators have been developing new techniques to find similar solutions. A recent impactful result, she says, was Brian Lawrence and Akshay

Venkatesh's new proof of something called Mordell's conjecture. Though Gerd Faltings first proved Mordell's conjecture in 1983, the work by Lawrence and Venkatesh "gives another perspective on a problem that's nearly 100 years old," Balakrishnan says. These and other advances show that interest in Diophantine geometry has been growing in recent years, Corwin says, "especially with the rise of new methods."

About the Author

Rachel Crowell is a Midwest-based writer covering science and mathematics.

New Proof Solves 80-Year-Old Irrational Number Problem

By Leila Sloman

Most people rarely deal with irrational numbers–it would be, well, irrational, as they run on forever, and representing them accurately requires an infinite amount of space. But irrational constants such as π and $\sqrt{2}$–numbers that cannot be reduced to a simple fraction–frequently crop up in science and engineering. These unwieldy numbers have plagued mathematicians since the ancient Greeks; indeed, legend has it that Hippasus was drowned for suggesting irrationals existed. Now, though, a nearly 80-year-old quandary about how well they can be approximated has been solved.

Many people conceptualize irrational numbers by rounding them to fractions or decimals: estimating π as 3.14, which is equivalent to 157/50, leads to widespread celebration of Pi Day on March 14th. Yet a different approximation, 22/7, is easier to wrangle and closer to π. This prompts the question: Is there a limit to how simple and accurate these approximations can ever get? And can we choose a fraction in any form we want?

In 1941 physicist Richard Duffin and mathematician Albert Schaeffer proposed a simple rule to answer these questions. Consider a quest to approximate various irrational numbers. First, decide on how close the approximation should be for fractions of a particular denominator. (Remember, the "numerator" refers to the top of a fraction and the "denominator" the bottom. Here, all the fractions are fully simplified–so, for example, 2/4 does not count as having the denominator 4 because it simplifies to 1/2.) You might decide that simplified fractions of the form $n/2$ can approximate any irrational number whose true value falls within 1/10 of them–giving the approximation an "error" of 1/10. Fractions that look like $n/10$ are closer together on the number line than those with denominator 2, so you might limit the error in that case to only 1/100–those fractions can approximate anything within 1/100th of them.

Usually, larger denominators are associated with smaller errors. If this is true, and there are infinitely many denominators that one can use to approximate a number to within the corresponding error, then by increasing the denominator the approximation can be made better and better. Duffin and Schaeffer's rule measures when this can be done based on the size of the errors.

If the chosen errors are small enough in aggregate, a randomly picked irrational number x will have only a limited number of good approximations: it might fall into the gaps between approximations with particular denominators. But if the errors are big enough, there will be infinitely many denominators that create a good approximating fraction. In this case, if the errors also shrink as the denominators get bigger, then you can choose an approximation that is as precise as you want.

Unproved

The upshot is that either you can approximate almost every number arbitrarily well, or almost none of them. "There's a striking dichotomy," says Dimitris Koukoulopoulos, a mathematician at the University of Montreal. Moreover, you can choose errors however you want, and as long as they are large enough in aggregate most numbers can be approximated infinitely many ways. This means that, by choosing some errors as zero, you can limit the approximations to specific types of fractions–for example, those with denominators that are powers of 10 only.

Although it seems logical that small errors make it harder to approximate numbers, Duffin and Schaeffer were unable to prove their conjecture–and neither was anybody else. The proof remained "a landmark open problem" in number theory, says Christoph Aistleitner, a mathematician at Graz University of Technology in Austria who has studied the problem. That is, until Koukoulopoulos and his co-author James Maynard announced their solution in a paper published in *Annals of Mathematics*.

The Duffin-Schaeffer conjecture "has this magical simplicity in an area of maths that's normally exceptionally difficult and complicated," Maynard says, a professor at the University of Oxford. He stumbled into the problem by accident–he is a number theorist, but not in the same area as most Duffin-Schaeffer experts. (He normally studies prime numbers–those that are divisible by only themselves and 1.) A University of York professor suggested Maynard tackle the Duffin-Schaeffer conjecture after he gave a talk there. "I think he had an intuition that it might be beneficial to get someone slightly outside of that immediate field," says Maynard. That intuition turned out to be correct, although it would not bear fruit for several years. Long after that initial conversation, Maynard suggested a collaboration to Koukoulopoulos on a suspicion that his colleague had relevant expertise.

Maynard and Koukoulopoulos knew that previous work in the field had reduced the problem to one about the prime factors of the denominators–the prime numbers that, when multiplied together, yield the denominator. Maynard suggested thinking about the problem as shading in numbers: "Imagine, on the number line, coloring in all the numbers close to fractions with denominator 100." The Duffin-Schaeffer conjecture says if the errors are large enough and one does this for every possible denominator, almost every number will be colored in infinitely many times.

For any particular denominator, only part of the number line will be colored in. If mathematicians could show that for each denominator, sufficiently different areas were colored, they would ensure almost every number was colored. If they could also prove those sections were overlapping, they could conclude that happened many times. One way of capturing this idea of different-but-overlapping areas is to prove the regions colored by different denominators had nothing to do with one another–they were independent.

But this is not actually true, especially if two denominators share many prime factors. For example, the possible denominators 10 and 100 share factors 2 and 5–and the numbers that can be approximated by fractions of the form $n/10$ exhibit frustrating overlaps with those that can be approximated by fractions $n/100$.

Graphing the Problem

Maynard and Koukoulopoulos solved this conundrum by reframing the problem in terms of networks that mathematicians call graphs–a bunch of dots, with some connected by lines (called edges). The dots in their graphs represented possible denominators that the researchers wanted to use for the approximating fraction, and two dots were connected by an edge if they had many prime factors in common. The graphs had a lot of edges precisely in cases where the allowed denominators had unwanted dependencies.

Using graphs allowed the two mathematicians to visualize the problem in a new way. "One of the biggest insights you need is to forget all the unimportant parts of the problem and to just home in on the one or two factors that make [it] very special," says Maynard. Using graphs, he says, "not only lets you prove the result, but it's really telling you something structural about what's going on in the problem." Maynard and Koukoulopoulos deduced that graphs with many edges corresponded to a particular, highly structured mathematical situation that they could analyze separately.

The duo's solution came as a surprise to many in the field. "The general feeling was that this was not close to being solved," says Aistleitner. "The technique of using [graphs] is something that maybe in the future will be regarded as just as important [as]–maybe more important than–the actual Duffin-Schaeffer conjecture," says Jeffrey Vaaler, a retired professor at the University of Texas, Austin, who proved a special case of the conjecture in 1978.

It may take other experts several months to understand the full details. "The proof now is a long and complicated proof," says Aistleitner. "It's not sufficient just to have one striking, brilliant idea.

There are many, many parts that have to be controlled." At 44 pages of dense, technical mathematics, even leading mathematical minds need time to wrap their heads around the paper. The community, however, seems optimistic. Says Vaaler: "It's a beautiful paper. I think it's correct."

About the Author

Leila Sloman received her PhD from Stanford University and is a Princeton-based science writer for Quanta, Scientific American, *and* Popular Mechanics.

Section 2: Society

Ask Gini: How to Measure Inequality

By Evelyn Lamb

Frank Cowell, an economist at the London School of Economics and Political Science, says the Gini coefficient is like the Kardashians: "It's famous for being famous." He's speaking about one of the most commonly discussed measures of income and wealth inequality. The Gini coefficient has been in the news a lot since the U.S. Census Bureau released its most recent data on income inequality in September. The data show that income inequality in the U.S. is high, but many articles blunder when they try to compare the U.S. to other countries.

An article on the *Atlantic* website in October, for example, reports on a pairing of U.S. cities with foreign countries that have similar Gini coefficients. The city in the U.S. with the least income inequality, Ogden, Utah, was paired with Malawi in Africa, whereas the city with the greatest inequality, Bridgeport, Conn., was paired with Thailand in Southeast Asia. These pairings are a bit puzzling. Is it better to be Malawi than Thailand? Does it make sense to compare the Gini coefficient of one concentrated metropolitan area with that of an entire nation?

A closer look at the data used to create the map shows that, as reported in a *Forbes* editorial, the U.S. Census Bureau usually reports Gini coefficients based on pretax numbers, whereas many calculations for foreign countries use post-tax numbers, which often include redistribution of wealth from rich to poor and tend to lower the Gini coefficient. Comparing the pretax number in one country with the post-tax number in another is somewhat meaningless.

To understand what the Gini coefficient can and cannot explain, and how to interpret articles about economic inequality, a deeper look at this statistic is required.

The Gini coefficient compares the income or wealth distribution of a population to a perfectly equal distribution–in which every citizen of a city or country has equal wealth. To compute the Gini coefficient, economists first find the Lorenz curve for the population. The curve is a graphical representation of the distribution of income or wealth in a society. The x-axis is the proportion of the population, from lowest to highest income, and the y-axis is the cumulative percentage of income or wealth owned. So the point (0.5, 0.2) would indicate that the lowest-income 50 percent of the population earned 20 percent of the total income. A perfectly equal society would have a Lorenz distribution that looks like the line $y = x$.

The Gini coefficient measures how far the actual Lorenz curve for a society's income or wealth is from the line of equality. Both the Lorenz curve and the line of equality are plotted on a graph. Then the area between the two graphs is computed. The Gini coefficient is the area between the two graphs divided by the total area under the line of inequality. This yields a number between 0 and 1, sometimes reported as a percentage—for example, 0.22 or 22 (written with or without the percent sign). 0 means that the country is perfectly equal, and 1 means that one person has all the wealth or income.

Cowell says that the Gini coefficient is useful, particularly because it allows negative values for income and wealth, unlike some other measures of inequality. (If some amount of the population has negative wealth (owes money), the Lorenz curve will dip below the x-axis.) But the Gini coefficient also has limitations. For one, it takes all the data from the Lorenz curve and converts it to a single number. Two different income distributions can have the same Gini coefficient, and a lot of information is lost in the conversion to a graph. Cowell asks, "Why not just look at the Lorenz curve?"

In addition, the Gini coefficient cannot tell that person X is a 24-year-old medical student who has negative income because of student loans, whereas person Y, who has the same amount of negative income, is unemployed and without job prospects. It samples people at random points of their lives, which means that it can't separate those whose financial futures are reasonably secure from those who do not have prospects. Its results are also sensitive to outliers—a few very wealthy or very poor individuals can change the statistic significantly, even in a large sample.

Cowell says that the Gini coefficient should not be used as the sole measure of economic inequality. He suggests two ways to handle the number: "One is to look beyond the Gini as a single statistic. The other is to consider whether it might be useful to use a model of the upper tail of the distribution, so you get a clearer picture." Due to incomplete data, the Gini coefficient can underestimate the concentration of wealth in the very richest individuals, and can even underestimate the wealth inequality within the upper echelons of the wealthy. To mitigate this problem, Cowell studies better ways to model income and wealth distribution in the most well-off. One option is to "patch in" an assumed distribution (specifically a Pareto distribution) for the top 5 or 10 percent of the population. In effect, this means assuming that the distribution of wealth takes a certain form, and using that model, rather than sparse data, to calculate the Gini coefficient.

The study of income and wealth inequality are of course fertile ground for many questions and controversies. What "should" the inequality in a society be? As the Occupy Wall Street (OWS) movement highlighted, this is of broad interest. Cowell says that the concentration of wealth in the upper echelon of the population can be reminiscent of a monopoly in business. "People get kind of twitchy if a large portion of the output is controlled by a small number of firms." Economists are asking many different questions about the causes and effects of wealth and income inequality. What are the effects of high inequality, and is it possible to separate the effects of poverty from the effects of inequality?

But the Gini coefficient is not just used by economists. Sam Shah, a high school math teacher in Brooklyn, N.Y., wrote in his blog that he included a section on the Gini coefficient during the last week of his calculus class. He framed the lessons in the context of OWS and asked, "Is income truly becoming more and more unequally distributed in the past 40 years?" Students had access to several decades' worth of data and got to explore questions about how they thought income and wealth should be distributed, in addition to working on math. As a way to motivate students' desire to understand calculus and statistics, he found the Gini coefficient to be very effective.

"The best part of the discussion was around what kids picked for 'what they would like it to be,'" wrote Shah. Others have studied this question: How do people feel about wealth inequality? A 2005 survey conducted by Michael Norton of the Harvard Business School and Dan Ariely of the Duke University Department of Psychology and published in 2011 found that most Americans underestimated the amount of wealth inequality in the U.S. and wanted it to be even lower than their estimates. The researchers showed respondents three different pie charts illustrating possible wealth distribution by quintiles. One illustrated complete equality, one was slightly unequal–with the lowest quintile earning 11 percent of the wealth and the highest earning 36 percent–and one was based on the wealth distribution of the U.S., with the lowest quintile owning 0.1 percent of the wealth and the top quintile owning 84 percent. Of the people surveyed, 47 percent preferred the slightly unequal distribution, 43 percent the perfectly equal distribution and only 10 percent the highly unequal distribution.

The diagram in the article labeled the slightly unequal distribution as Sweden, although it was presented without a label to survey respondents. The authors clearly wanted readers to believe that Sweden's wealth distribution was preferable to that of the U.S.–a heading in the article stated "Americans Prefer Sweden." But in a note at the end of the article, the authors wrote, "We used Sweden's income rather than wealth distribution because it provided a clearer contrast to the other two wealth distribution examples; although more equal than the United States's wealth distribution, Sweden's wealth distribution is still extremely top-heavy." (According to Cowell's research, even that statement is unclear: some methods of computing the Gini coefficient that include a modification of the distribution at the top of the wealth scale find that Sweden has greater wealth inequality than the U.S.)

Although Norton and Ariely's conclusion that Americans would prefer a more equal distribution of wealth may be sound, their data switcheroo illustrates a common problem when talking about inequality: income versus wealth. People often conflate the two, but they are not the same. "You're using your thermometer to measure something quite different," Cowell says. Wealth inequality says more about the balance of power in a society, and income inequality addresses the way labor markets operate. "Typically wealth is much more unequally distributed than income, in any country you look at," he says. He recently conducted a study about income and wealth inequality in the U.S., U.K., Canada and Sweden. Because Sweden has more interventionist policies, one might assume that the U.S. would be much more unequal than Sweden. "It's true for income but not for wealth," says Cowell.

Economists continue to use the Gini coefficient, either standard or modified, to understand wealth and income inequality. In the meantime, lay people who want to understand wealth inequality should read the fine print to ensure that they have all the facts.

About the Author

Evelyn Lamb is a freelance math and science writer based in Salt Lake City, Utah.

Can an Algorithm Help Solve Political Paralysis?

By Julia Hotz

Dave Johnson refuses to eat fish farmed near Minneapolis. The retired construction worker, who lives in a rural area 150 miles north of the Twin Cities, claims "tourists' garbage, salt runoff and fertilizer" make the metropolitan area's farmed fish inedible. But for Johnson, the "divide between city people and rural people" is about much more than fish: it's about how individuals relate—or don't relate—to the environment. "People in cities are busy living in the fast lane, while rural communities are really attuned to what's going on in nature," explains Johnson, who says he voted for Donald Trump in 2016 and links the country's bitter urban-rural divide to bad policy making: "A lot of the politicians at the state capital pass laws because there's peer pressure," he says. "But they don't really know what life is like up here—and they don't care about us."

Shona Snater cites a different kind of divide. Instead of geography, the 31-year-old soil-health organizer in southeast Minnesota says "corporate interests [are] dictating how politicians think and act," especially when it comes to the environment. "There's progress at the grassroots level," she adds. "But when farmers are using half of the fertilizer they used to use, that's terrifying for agribusinesses." A registered Democrat, Snater says she believes both of the dominant U.S. political parties can be "bought out." She blames corporations for supporting "terrible policies" that favor "economies of scale" and leave many people behind. "Small and midsize farmers say, 'These politicians aren't representing my best interests,'" Snater says, noting that Minnesota has been losing dairy farms at a rate of nearly one a day.

Such frustrations around the usual mechanics of policy making are rife throughout much of the U.S. In a 2018 Pew Research Center survey of 10,000-plus Americans, 75 percent said trust in the federal government has been shrinking. And when the survey asked *who*

Americans trusted to "act in the best interests of the public," the majority of respondents revealed they had the most confidence in scientists and the military–and the least in elected officials. Brett Hennig, an advocate for alternative ways of conducting democracy, says he can understand why.

"If you believe an ideal democracy involves informed deliberation among a representative group of people, the current electoral system fails on both counts," says Hennig, who has a Ph.D. in astrophysics. He believes something called "citizens' assemblies" offer a better way to elicit policies in line with people's real interests–with a little help from an algorithm.

Hennig explains citizens' assemblies using simple logic: society is made up of people who are young and old, rich and poor, and mostly in between, so decisions governing it should more directly involve a group proportionally representing these kinds of characteristics. But because many ordinary citizens may lack technical knowledge of the issues at hand, citizens' assemblies invite these individuals to make decisions in a "deliberative environment"–in which they can consult experts to "reduce the effect of biases, misleading information and ignorance" when learning about a problem and assessing possible solutions, Hennig says. From there, these citizens collaboratively craft recommendations for policy makers to consider.

Philipp Verpoort, a Ph.D. candidate in physics at the University of Cambridge in England, is another scientist who advocates for citizens' assemblies. "Everyone's talking about the three P's: pessimism, populism and polarization. And we're at this point where people split up into groups, don't trust their politicians, and nothing gets done," he says. "But when people see a decision being made by people like them, they trust it."

To turn their theories to practice, Hennig, Verpoort and their colleagues co-direct the Sortition Foundation–a nonprofit organization that offers "selection and stratification services" for citizens' assemblies and similar deliberative bodies. The foundation (which is financed largely by payments received for its services) has supported about 20 such projects among the roughly 200 that various organizing bodies have hosted around the world, according to Verpoort and Hennig. One of the most famous efforts, which the Sortition Foundation was not involved in, was held in Ireland, where a 99-person assembly prompted the nation in 2018 to repeal a law that had effectively banned abortion. But much of the buzz today surrounds climate change and the U.K.: after a series of smaller city-based gatherings in 2019, 110 citizens were selected earlier this year for the country's Climate Assembly UK project. Through this deliberation, participants were invited to recommend ways the U.K. government could meet its legally binding target of net-zero greenhouse gas emissions by 2050.

After meeting in person and virtually over six weekends between January and May, the participants recently published their final report of policy recommendations, which range from levying a tax on frequent air travelers to investing in low-carbon public transportation.

Using an Algorithm to Build a Scale Model of a Society

With a mathematician's precision, Hennig explains how an algorithm he created generated the 110-person "mini public" from the U.K.'s population of 67 million. The process began by sending invitations to 30,000 households from the nation's postal database. Hennig says completely random selection would have skewed the responses toward people with higher incomes (who are more likely to have the time and money to participate). So 20 percent of the sampled individuals were randomly invited from the "most deprived areas," and 80 percent were chosen at random from every region. To further reduce the effects of income-related selection bias, participants were promised a small stipend and travel reimbursements.

Out of the 30,000 people invited, nearly 2,000 accepted and completed an online survey indicating seven characteristics: their gender identity, age, ethnicity, educational attainment, location, description of their residence as urban or rural, and level of concern about climate change. Hennig applied his algorithm to those

2,000 respondents to select 110 participants who would proportionally represent the U.K. with respect to those seven categories.

To begin the process, the algorithm went through the survey answers and randomly and repeatedly selected from the "hardest category to fill," Hennig says. He adds that it was initially (and unsurprisingly) hardest to find respondents among those who reported they were "not very concerned" or "not at all concerned" about climate change. The algorithm continued by selecting for the "most in need" demographic–"determined by the maximum of the ratio between number of people still required for fill a category quota and the number of people available for selection in that category," Verpoort explains–until it drew 110 names. In the end, though the Climate Assembly UK website cautions that the algorithm "slightly over sampled" certain demographics, the selected group's distribution of the seven categories very closely matched that of the U.K.

But Jonathan Mattingly, a Duke University professor of mathematics who is also using algorithms in an attempt to help fix democracy, takes issue with that framework. "Who's determining which categories inform decision-making? Maybe it's your family income before you turned 16, or whether your parents were immigrants," Mattingly says, referring to the seven characteristics used by Hennig and Verpoort. "How do we know which factors account for clarity, empathy and decisiveness?"

Geoff Bedford, a participant in a "citizens' jury" (a smaller kind of citizens' assembly) on climate, in the city of Leeds, England, also expresses concerns about how such assemblies are run. He points specifically to the expertise process. "Most [participants] seemed to rely solely on the opinion of so-called climate scientists," says Bedford, who doubts that humans cause climate change. The participants, he adds, "don't have time to read the conclusions of [the experts'] reports and, understandably, accept sound bites from the mainstream media."

Verpoort and Hennig counter that, for practicality's sake, most assembly organizers prioritize experts whose research is peer-reviewed–and the vast majority of such research confirms anthropogenic global warming. But to make sure people like Bedford were still a part of the conversation, another algorithm was used during Climate Assembly UK. "Once you've got a diverse group of people representing society together in a room, you don't want all of the climate skeptics to end up on the same table and talk to nobody else," Verpoort says. To avoid this clustering, he created an algorithm to divide the 110-person sample into tables of eight individuals, among whom the seven categories–age, location, and so on–were proportionally represented.

Can Citizens' Assemblies Work in the U.S.?

Even if citizens' assemblies prove effective elsewhere, some doubt they would work as well in the U.S.–and that the country's federal and local governments would take the resulting recommendations very seriously. But Crystal Chissell, vice President of operations and engagement at the nonprofit environmental organization Project Drawdown, says she is "absolutely" convinced that ordinary Americans can help craft actionable policies. She points to Cincinnati, where hundreds of residents participated in developing the Green Cincinnati Plan of 80 evidence-based strategies to reduce the city's carbon emissions by 80 percent by 2050. "'Ordinary people' are very savvy when it comes to seeing the co-benefits of climate solutions," Chissell says, explaining that many simply want policies benefiting their health, well-being and finances while also addressing climate change.

One of the U.S.'s prominent champions of people-powered policy making is James Fishkin, a professor of communication and political science at Stanford University. He is known for developing Deliberative Polling, a process that–like citizens' assemblies–uses stratified random sampling to gather a representative group of citizens, facilitate informed deliberation among them, and highlight "actionable priorities" for governments.

In the past, Fishkin's polls have been used to advance policies such as sustainable energy choices in Texas, Nebraska and Vermont.

And he says that America in One Room, a recent national gathering of 523 citizens (a stratified, random sample that was representative of the American electorate in terms of attitudes and demographics) showed they can also build understanding across political ideologies. "When a candidate wants to win an election, they deploy one-sided arguments just to generate outrage," Fishkin says. "But when you empower people to go beyond tribalism and consider an issue under the best possible conditions–in an evidence-based discussion with multiple perspectives–there's more policy consensus and less gravitation toward extremes."

Johnson, the retired construction worker, may offer some proof of that theory. He recently attended the Rural Climate Dialogues, a series of Minnesota-based "citizens' juries" that were hosted by the nonprofit Jefferson Center, which was one of the earliest adopters of such juries in the U.S. After taking part, he says he now firmly disagrees with Trump's position on climate change and better understands its effects. "To hear how climate change is behind polar bears losing their homes, our changing forests and the heavy rains that keep my construction buddies from working–I just learned so much," Johnson adds.

Snater, the soil-health organizer, also says the Rural Climate Dialogues opened her mind to new perspectives. Although she says she has spent much of her life researching climate change, Snater believes the jury helped her better understand how the phenomenon relates to the on the ground experiences of her neighbors. "This woman [from the dialogues] told a story about running out of fuel and living in a cold house for a couple of weeks with her kids," she says. "It was a real moment where I realized, 'Not everyone is in the same position I am.'"

About the Author

Julia Hotz is a journalist reporting on systemic solutions to loneliness, anxiety, depression and disease.

Simple Mathematical Law Predicts Movement in Cities around the World

By Viviane Callier

The people who happen to be in a city center at any given moment may seem like a random collection of individuals. But new research featuring a simple mathematical law shows that urban travel patterns worldwide are, in fact, remarkably predictable regardless of location—an insight that could enhance models of disease spread and help to optimize city planning.

Studying anonymized cell-phone data, researchers discovered what is known as an inverse square relation between the number of people in a given urban location and the distance they traveled to get there, as well as how frequently they made the trip. It may seem intuitive that people visit nearby locations frequently and distant ones less so, but the newly discovered relation puts the concept into specific numerical terms. It accurately predicts, for instance, that the number of people coming from two kilometers away five times per week will be the same as the number coming from five kilometers twice a week. The researchers' new visitation law, and a versatile model of individuals' movements within cities based on it, was reported in *Nature*.

"This is a super striking, robust result," says Laura Alessandretti, a computational social scientist at the Technical University of Denmark, who was not involved in the study but co-wrote an accompanying commentary. "We tend to think that there are lots of contextual aspects that affect the way we move, such as the transportation system, the morphology of a given place, and socioeconomic aspects. This is true to some extent, but what this shows is that there are some robust laws that apply everywhere."

The researchers analyzed data from about eight million people between 2006 and 2013 in six urban locations: Boston, Singapore, Lisbon and Porto in Portugal, Dakar in Senegal, and Abidjan in

Ivory Coast. Previous analyses have used cell-phone data to study individuals' travel paths; this study focused instead on locations and examined how many people were visiting, from how far and how frequently. The researchers found that all the unique choices people make–from dropping kids at school to shopping or commuting–obey this inverse square law when considered in aggregate. "The result is very simple but quite startling," says Geoffrey West, an urban scaling theorist at the Santa Fe Institute and one of the paper's senior authors.

One explanation for this strong statistical pattern is that traveling requires time and energy, and people have limited resources for it. "There is something really very fundamental at play here. Whether you live in Senegal or in Boston, you try to optimize your day," says study lead author Markus Schläpfer of ETH Zurich's Future Cities Laboratory in Singapore. "At the core is the effort that people are willing to invest collectively to travel to certain locations."

Understanding these patterns is important not only for planning the placement of new shopping centers or public transportation but also for modeling disease transmission within cities, says Kathleen Stewart, a geographer and mobility researcher at the University of Maryland who was not involved in the study.

Many researchers estimate travel with "gravity models," which assume that movement between cities is proportional to their population sizes. But these models do not account for travel patterns within cities–information that is particularly critical in tackling disease transmission. Northeastern University epidemiologist Sam Scarpino, who was not involved in the study, says models based on this new finding might better track that flow. For example, New York City residents are more likely to make short, frequent trips within their own borough (such as Manhattan or the Bronx) and fewer trips to a distant borough.

"Those organizational patterns have really profound implications on how COVID will spread," Scarpino says. In a smaller rural location, where many people regularly go to the same church or grocery store, the entire town will experience sharp peaks of infections as the virus sweeps through the community. But in a bigger city, the propagation takes longer, he explains, because mini epidemics can occur in each neighborhood somewhat separately.

Stewart adds: "The authors demonstrate that their visitation law–that takes into account both travel distance and frequency of visits in a way that other models do not–outperforms gravity models when it comes to predicting flows between locations."

About the Author

Viviane Callier is a freelance science writer based in North Bethesda, Md.

Smartphone Data Show Voters in Black Neighborhoods Wait Longer

By Daniel Garisto

On November 8, 2016, tens of million of Americans went to vote at about 110,000 polling places. Most were in and out in under 10 minutes, but many still waited in line a long while–in some cases, for hours. Until now, a comprehensive nationwide account of how much time voters spent waiting was out of reach.

In a study led by economist Keith Chen of the University of California, Los Angeles, researchers matched anonymous location data from 10 million smartphones to 93,000 polling places to create the most extensive map to date of voter wait times across the U.S. The results, reported in January 2020 by the National Bureau of Economic Research, showed one very clear disparity: voters in predominantly black neighborhoods waited 29 percent longer, on average, than those in white neighborhoods. They were also about 74 percent more likely to wait for more than half an hour.

This study "was a totally, totally different way to try to measure this problem than what we've seen before–and it comes to the same conclusion," says Stephen Pettigrew of the University of Pennsylvania, an expert in voting wait time, who was not involved with the study.

Timing the Vote

Long lines at polling places have made headline after headline in recent years, and have been the subject of a flurry of research by political scientists. Many see protracted voting wait times not only as an inconvenience but as a civil rights issue.

"[It] really creates a barrier to the most vulnerable voters out there," says Sophia Lin Lakin, an attorney at the American Civil Liberties Union (ACLU). Minority voters often have less flexible work hours, she says, so lengthy wait times can reduce their ability to vote. Long lines are estimated to have deterred between 500,000 to 700,000 people from casting their ballot in 2012. These problems led to the creation of the Presidential Commission on Election Administration, which issued a 2014 report that set forth a standard: "No citizen should have to wait more than 30 minutes to vote."

Measuring how much time voters really spend waiting, though, is easier said than done. Previous studies have typically relied on two methods: self-reported surveys and in-person poll observers. Studies using these methods generally agreed on a basic result: in 2012 and 2016 the average nonwhite voter waited about twice as long as the average white voter. Poll observers can precisely track an individual's wait time, and surveys are generally quite accurate. But both methods suffer from limited reach; the most comprehensive prior study covered only 528 polling places.

This data gap is where the new method shines: Chen and his colleagues collected addresses for 93,658 polling places–80 percent of the U.S. total–and converted each into latitude and longitude coordinates, effectively creating a map of voting spots across the country.

Smartphones "ping," or send out data with their location, every few minutes on average. Using proprietary data from SafeGraph, a company that collects such information from smartphone apps, the researchers gathered pings that came from within 60 meters of a polling place during the 2016 presidential election. From this sample, they excluded people unlikely to be voters: for example, those who were in the area all day and thus likely to be poll workers, and those who were there for less than one minute and thus likely to be just passing by. After doing so, they ended up with a sample of more than 150,000 voters at around 40,000 polling locations. Using demographic data from the U.S. Census, Chen and his colleagues compared the wait times of voters by neighborhood–and found those in majority-black neighborhoods (as well as other non-white-majority neighborhoods) waited longer.

Although the wait times the researchers calculated are not as precise as past studies, which could identify individual voters,

Pettigrew says the results are a trustworthy indicator of this trend. By 2016 two thirds of all Americans owned a smartphone, and voters who did not have one would presumably have been just as likely to get stuck in line. The study contended that its method was, in fact, biased toward not finding a racial disparity in wait times: false positives–nonvoters who resembled voters, based on smartphone data–would have acted like random noise and should have made it hard to see an effect. But the difference between black and white neighborhoods remains clear.

Still, the smartphone method does not answer a key question: "What it doesn't tell us is *why*. What's going on here?" says Robert Stein, a political scientist at Rice University, who did not participate in the new research. "I would want to use it as a basis for doing further observational studies. You know, you just can't get that from the phone."

In the new study, the researchers acknowledged the phone data were only a proxy for voting wait time, which could result from anything–ballots that take a while to fill out, a lack of parking spaces or too few voting machines. "It basically tells us who's within 200 feet [60 meters] of a polling place on Election Day," says Charles Stewart, a political scientist at the Massachusetts Institute of Technology, who was also not involved in the research.

Suppressing Turnout

Unlike explicit methods of voter suppression, such as Jim Crow–era laws, long voting wait times do not necessarily result from a direct intent to discriminate. Previous studies have found that some areas–often those that are nonwhite–lack the resources to properly accommodate their voters.

But in a 2016 study, Stein and his colleagues did find one particular culprit for long lines: voter ID laws. In many states, voters who do not have a certain form of identification must fill out a provisional ballot, a time-consuming ordeal that can clog up a polling place. Like cars on a highway, even a single stopped voter can slow down the whole line. Areas with minority voters–who are less likely to have an ID–tend to be most affected by these laws, Stein says.

The past decade has seen numerous challenges to voting, including gerrymandering (manipulating districts in a way that politically disenfranchises some voters), voter-roll purges (challenging people's registration status to keep them from voting), and, in 2016, targeted disinformation campaigns to suppress black voters. "Significantly, the Supreme Court has neutered the core protection of the federal Voting Rights Act," says Wendy Weiser, director of the Democracy Program at the Brennan Center for Justice at New York University. Previously, states with a history of discriminatory voting laws had to receive clearance from the federal government before making any changes to voting practices. But since the court's 2013 decision, nearly 1,700 polling places–many in black and Latino neighborhoods–have been shuttered. In 2017 the ACLU of Georgia sued a county elections board in the state for closing several polling places without adequate notice (more than 200 voting precincts have been closed in Georgia since 2012). "From our perspective, that's an unconstitutional burden on these individuals' right to vote," Lakin says.

With congressional elections every two years and presidential elections every four, the authors of the study wrote that their method is an "easily available and repeatable tool to both diagnose and monitor" voting wait times. But getting these kinds of smartphone data is difficult because they are proprietary, and getting the locations of nearly 100,000 polling places is no mean feat. "It's an incredibly rich data set, and there would be huge opportunities for either [the study authors] or for other researchers to take what they have and build on it," Pettigrew says. "Putting the data out there would be a huge step in the right direction."

About the Author

Daniel Garisto is a freelance science journalist covering advances in physics and other natural sciences. His writing has appeared in Nature News, Science News, Undark, *and elsewhere.*

The American Economy Is Rigged

By Joseph E. Stiglitz

Americans are used to thinking that their nation is special. In many ways, it is: the U.S. has by far the most Nobel Prize winners, the largest defense expenditures (almost equal to the next 10 or so countries put together) and the most billionaires (twice as many as China, the closest competitor). But some examples of American Exceptionalism should not make us proud. By most accounts, the U.S. has the highest level of economic inequality among developed countries. It has the world's greatest per capita health expenditures yet the lowest life expectancy among comparable countries. It is also one of a few developed countries jostling for the dubious distinction of having the lowest measures of equality of opportunity.

The notion of the American Dream–that, unlike old Europe, we are a land of opportunity–is part of our essence. Yet the numbers say otherwise. The life prospects of a young American depend more on the income and education of his or her parents than in almost any other advanced country. When poor-boy-makes-good anecdotes get passed around in the media, that is precisely because such stories are so rare.

Things appear to be getting worse, partly as a result of forces, such as technology and globalization, that seem beyond our control, but most disturbingly because of those within our command. It is not the laws of nature that have led to this dire situation: it is the laws of humankind. Markets do not exist in a vacuum: they are shaped by rules and regulations, which can be designed to favor one group over another. President Donald Trump was right in saying that the system is rigged–by those in the inherited plutocracy of which he himself is a member. And he is making it much, much worse.

America has long outdone others in its level of inequality, but in the past 40 years it has reached new heights. Whereas the income share of the top 0.1 percent has more than quadrupled and that of the top 1 percent has almost doubled, that of the bottom 90 percent has declined. Wages at the bottom, adjusted for inflation, are about the same as they were some 60 years ago! In fact, for those with a high school education or less, incomes have fallen over recent decades. Males have been particularly hard hit, as the U.S. has moved away from manufacturing industries into an economy based on services.

Deaths of Despair

Wealth is even less equally distributed, with just three Americans having as much as the bottom 50 percent–testimony to how much money there is at the top and how little there is at the bottom. Families in the bottom 50 percent hardly have the cash reserves to meet an emergency. Newspapers are replete with stories of those for whom the breakdown of a car or an illness starts a downward spiral from which they never recover.

In significant part because of high inequality, U.S. life expectancy, exceptionally low to begin with, is experiencing sustained declines. This in spite of the marvels of medical science, many advances of which occur right here in America and which are made readily available to the rich. Economist Ann Case and 2015 Nobel laureate in economics Angus Deaton describe one of the main causes of rising morbidity–the increase in alcoholism, drug overdoses and suicides–as "deaths of despair" by those who have given up hope.

Defenders of America's inequality have a pat explanation. They refer to the workings of a competitive market, where the laws of supply and demand determine wages, prices and even interest rates–a mechanical system, much like that describing the physical universe. Those with scarce assets or skills are amply rewarded, they argue, because of the larger contributions they make to the economy. What they get merely represents what they have contributed. Often they take out less than they contributed, so what is left over for the rest is that much more.

This fictional narrative may at one time have assuaged the guilt of those at the top and persuaded everyone else to accept this sorry state of affairs. Perhaps the defining moment exposing the lie was the 2008 financial crisis, when the bankers who brought the global economy to the brink of ruin with predatory lending, market manipulation and various other antisocial practices walked away with millions of dollars in bonuses just as millions of Americans lost their jobs and homes and tens of millions more worldwide suffered on their account. Virtually none of these bankers were ever held to account for their misdeeds.

I became aware of the fantastical nature of this narrative as a schoolboy, when I thought of the wealth of the plantation owners, built on the backs of slaves. At the time of the Civil War, the market value of the slaves in the South was approximately half of the region's total wealth, including the value of the land and the physical capital—the factories and equipment. The wealth of at least this part of this nation was not based on industry, innovation and commerce but rather on exploitation. Today we have replaced this open exploitation with more insidious forms, which have intensified since the Reagan-Thatcher revolution of the 1980s. This exploitation, I will argue, is largely to blame for the escalating inequality in the U.S.

After the New Deal of the 1930s, American inequality went into decline. By the 1950s inequality had receded to such an extent that another Nobel laureate in economics, Simon Kuznets, formulated what came to be called Kuznets's law. In the early stages of development, as some parts of a country seize new opportunities, inequalities grow, he postulated; in the later stages, they shrink. The theory long fit the data—but then, around the early 1980s, the trend abruptly reversed.

Explaining Inequality

Economists have put forward a range of explanations for why inequality has in fact been increasing in many developed countries. Some argue that advances in technology have spurred the demand for skilled labor relative to unskilled labor, thereby depressing the wages of the latter. Yet that alone cannot explain why even skilled labor has done so poorly over the past two decades, why average wages have done so badly and why matters are so much worse in the U.S. than in other developed nations. Changes in technology are global and should affect all advanced economies in the same way. Other economists blame globalization itself, which has weakened the power of workers. Firms can and do move abroad unless demands for higher wages are curtailed. But again, globalization has been integral to all advanced economies. Why is its impact so much worse in the U.S.?

The shift from a manufacturing to a service-based economy is partly to blame. At its extreme–a firm of one person–the service economy is a winner-takes-all system. A movie star makes millions, for example, whereas most actors make a pittance. Overall, wages are likely to be far more widely dispersed in a service economy than in one based on manufacturing, so the transition contributes to greater inequality. This fact does not explain, however, why the average wage has not improved for decades. Moreover, the shift to the service sector is happening in most other advanced countries: Why are matters so much worse in the U.S.?

Again, because services are often provided locally, firms have more market power: the ability to raise prices above what would prevail in a competitive market. A small town in rural America may have only one authorized Toyota repair shop, which virtually every Toyota owner is forced to patronize. The providers of these local services can raise prices over costs, increasing their profits and the share of income going to owners and managers. This, too, increases inequality. But again, why is U.S. inequality practically unique?

In his celebrated 2013 treatise *Capital in the Twenty-First Century*, French economist Thomas Piketty shifts the gaze to capitalists. He suggests that the few who own much of a country's capital save so much that, given the stable and high return to capital (relative to the growth rate of the economy), their share of the national income has been increasing. His theory has, however, been questioned on many grounds. For instance, the savings rate of even the rich in the U.S. is so low, compared with the rich in other countries, that the increase in inequality should be lower here, not greater.

An alternative theory is far more consonant with the facts. Since the mid-1970s the rules of the economic game have been rewritten, both globally and nationally, in ways that advantage the rich and disadvantage the rest. And they have been rewritten further in this perverse direction in the U.S. than in other developed countries–even though the rules in the U.S. were already less favorable to workers. From this perspective, increasing inequality is a matter of choice: a consequence of our policies, laws and regulations.

In the U.S., the market power of large corporations, which was greater than in most other advanced countries to begin with, has increased even more than elsewhere. On the other hand, the market power of workers, which started out less than in most other advanced countries, has fallen further than elsewhere. This is not only because of the shift to a service-sector economy–it is because of the rigged rules of the game, rules set in a political system that is itself rigged through gerrymandering, voter suppression and the influence of money. A vicious spiral has formed: economic inequality translates into political inequality, which leads to rules that favor the wealthy, which in turn reinforces economic inequality.

Feedback Loop

Political scientists have documented the ways in which money influences politics in certain political systems, converting higher economic inequality into greater political inequality. Political inequality, in its turn, gives rise to more economic inequality as the rich use their political power to shape the rules of the game in ways that favor them–for instance, by softening antitrust laws and weakening unions. Using mathematical models, economists such as myself have shown that this two-way feedback loop between money and regulations leads to at least two stable points. If an economy starts out with lower inequality, the political system generates rules that sustain it, leading to one equilibrium situation. The American system is the other equilibrium–and will continue to be unless there is a democratic political awakening.

An account of how the rules have been shaped must begin with antitrust laws, first enacted in the1890s in the U.S. to prevent the agglomeration of market power. Their enforcement has weakened–at a time when, if anything, the laws themselves should have been strengthened. Technological changes have concentrated market power in the hands of a few global players, in part because of so-called network effects: you are far more likely to join a particular social network or use a certain word processor if everyone you know is already using it. Once established, a firm such as Facebook or Microsoft is hard to dislodge. Moreover, fixed costs, such as that of developing a piece of software, have increased as compared with marginal costs–that of duplicating the software. A new entrant has to bear all these fixed costs up front, and if it does enter, the rich incumbent can respond by lowering prices drastically. The cost of making an additional e-book or photo-editing program is essentially zero.

In short, entry is hard and risky, which gives established firms with deep war chests enormous power to crush competitors and ultimately raise prices. Making matters worse, U.S. firms have been innovative not only in the products they make but in thinking of ways to extend and amplify their market power. The European Commission has imposed fines of billions of dollars on Microsoft and Google and ordered them to stop their anti-competitive practices (such as Google privileging its own comparison shopping service). In the U.S., we have done too little to control concentrations of market power, so it is not a surprise that it has increased in many sectors.

Rigged rules also explain why the impact of globalization may have been worse in the United States. A concerted attack on unions has almost halved the fraction of unionized workers in the nation, to about 11 percent. (In Scandinavia, it is roughly 70 percent.) Weaker unions provide workers less protection against the efforts of firms to drive down wages or worsen working conditions. Moreover, U.S. investment treaties such as the North Atlantic Free Trade Agreement–treaties that were sold as a way of preventing foreign countries from discriminating against American firms–also protect investors against a tightening of environmental and health regulations abroad. For instance, they enable corporations to sue nations in private international arbitration panels for passing laws that protect citizens and the environment but threaten the multinational company's bottom line. Firms like these provisions, which enhance the credibility of a company's threat to move abroad if workers do not temper their demands. In short, these investment agreements weaken U.S. workers' bargaining power even further.

Global Inequality Trends

Inequality has increased in most advanced countries because of factors such as globalization, technological change, and the shift to a service-based economy. It has grown fastest in the U.S., however, according to the World Inequality Database. That is because rules have been rewritten to make them more favorable for the rich, while being disadvantageous to everyone else. Large companies have been allowed to accrue more power over the market, whereas the influence of workers has shrunk. Taxation and other policies have consistently favored the wealthy.

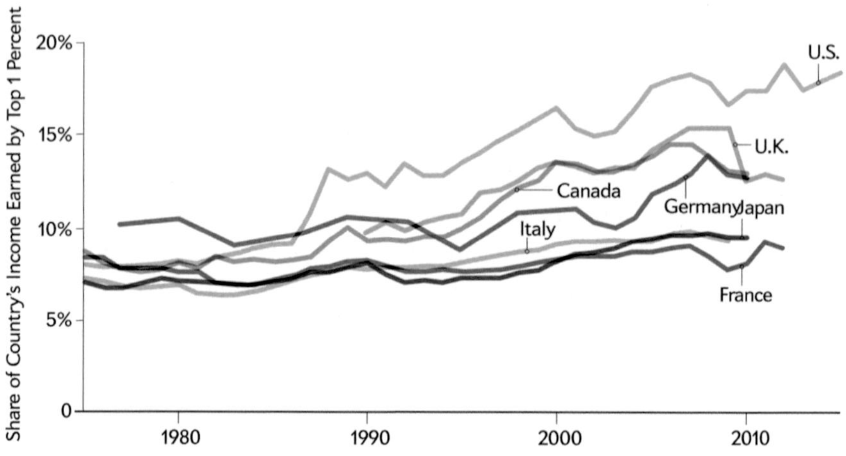

Credit: Jen Christiansen: Sources: *Economic Report of the President*. January 2017; World Inequality database.

Liberated Finance

Many other changes to our norms, laws, rules and regulations have contributed to inequality. Weak corporate governance laws have allowed chief executives in the U.S. to compensate themselves 361 times more than the average worker, far more than in other developed countries. Financial liberalization–the stripping away of regulations designed to prevent the financial sector from imposing harms, such as the 2008 economic crisis, on the rest of society–has enabled the finance industry to grow in size and profitability and has increased its opportunities to exploit everyone else. Banks routinely indulge in practices that are legal but should not be, such as imposing usurious interest rates on borrowers or exorbitant fees on merchants for credit and debit cards and creating securities that are designed to fail. They also frequently do things that are illegal, including market manipulation and insider trading. In all of this, the financial sector has moved money away from ordinary Americans to rich bankers and the banks' shareholders. This redistribution of wealth is an important contributor to American inequality.

Other means of so-called rent extraction–the withdrawal of income from the national pie that is incommensurate with societal contribution–abound. For example, a legal provision enacted in 2003 prohibited the government from negotiating drug prices for Medicare–a gift of some $50 billion a year or more to the pharmaceutical industry. Special favors, such as extractive industries' obtaining public resources such as oil at below fair-market value or banks' getting funds from the Federal Reserve at near-zero interest rates (which they re-lend at high interest rates), also amount to rent extraction. Further exacerbating inequality is favorable tax treatment for the rich. In the U.S., those at the top pay a smaller fraction of their income in taxes than those who are much poorer–a form of largesse that the Trump administration has just worsened with the 2017 tax bill.

Some economists have argued that we can lessen inequality only by giving up on growth and efficiency. But recent research, such as work done by Jonathan Ostry and others at the International Monetary Fund, suggests that economies with greater equality perform better, with higher growth, better average standards of living and greater stability. Inequality in the extremes observed in the U.S. and in the manner generated there actually damages the economy. The exploitation of market power and the variety of other distortions I have described, for instance, makes markets less efficient, leading to underproduction of valuable goods such as basic research and overproduction of others, such as exploitative financial products.

Moreover, because the rich typically spend a smaller fraction of their income on consumption than the poor, total or "aggregate" demand in countries with higher inequality is weaker. Societies could make up for this gap by increasing government spending–on infra-structure, education and health, for instance, all of which are investments necessary for long-term growth. But the politics of unequal societies typically puts the burden on monetary policy: interest rates are lowered to stimulate spending. Artificially low interest rates, especially if coupled with inadequate financial market regulation, can give rise to bubbles, which is what happened with the 2008 housing crisis.

It is no surprise that, on average, people living in unequal societies have less equality of opportunity: those at the bottom never get the education that would enable them to live up to their potential. This fact, in turn, exacerbates inequality while wasting the country's most valuable resource: Americans themselves.

Uneven Distribution of Global Growth

Globalization has benefited millions of the poor in emerging economies, particularly in China. Data compiled by economist Branko Milanovic and displayed in the World Inequality Report 2018 demonstrate, however, that between 1980 and 2016, the steepest gains went to the world's top 1 percent, which captured more than a quarter of the growth in the global economy. In early

2018 Oxfam International reported that just 42 individuals have as much wealth as the bottom 50 percent put together. The middle classes in the U.S. and western Europe have benefited the least from global growth, as have the world's poorest.

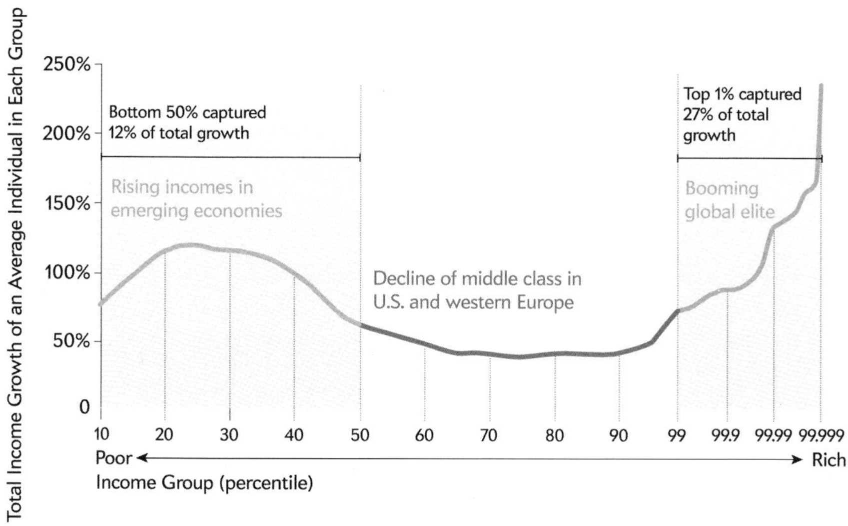

Credit: Jen Christiansen: Sources: *World Inequality Report 2018.* World Inequality Lab, 2017; Branko Milanovic.

Restoring Justice

Morale is lower in unequal societies, especially when inequality is seen as unjust, and the feeling of being used or cheated leads to lower productivity. When those who run gambling casinos or bankers suffering from moral turpitude make a zillion times more than the scientists and inventors who brought us lasers, transistors and an understanding of DNA, it is clear that something is wrong. Then again, the children of the rich come to think of themselves as a class apart, entitled to their good fortune, and accordingly more likely to break the rules necessary for making society function. All of this contributes to a breakdown of trust, with its attendant impact on social cohesion and economic performance.

There is no magic bullet to remedy a problem as deep-rooted as America's inequality. Its origins are largely political, so it is hard to imagine meaningful change without a concerted effort to take money out of politics–through, for instance, campaign finance reform. Blocking the revolving doors by which regulators and other government officials come from and return to the same industries they regulate and work with is also essential.

Beyond that, we need more progressive taxation and high-quality federally funded public education, including affordable access to universities for all, no ruinous loans required. We need modern competition laws to deal with the problems posed by 21st-century market power and stronger enforcement of the laws we do have. We need labor laws that protect workers and their rights to unionize. We need corporate governance laws that curb exorbitant salaries bestowed on chief executives, and we need stronger financial regulations that will prevent banks from engaging in the exploitative practices that have become their hallmark. We need better enforcement of anti-discrimination laws: it is unconscionable that women and minorities get paid a mere fraction of what their white male counterparts receive. We also need more sensible inheritance laws that will reduce the intergenerational transmission of advantage and disadvantage.

The basic perquisites of a middle-class life, including a secure old age, are no longer attainable for most Americans. We need to guarantee access to health care. We need to strengthen and reform retirement programs, which have put an increasing burden of risk management on workers (who are expected to manage their portfolios to guard simultaneously against the risks of inflation and market collapse) and opened them up to exploitation by our financial sector (which sells them products designed to maximize bank fees rather than retirement security). Our mortgage system was our Achilles' heel, and we have not really fixed it. With such a large fraction of Americans living in cities, we have to have urban housing policies that ensure affordable housing for all.

It is a long agenda–but a doable one. When skeptics say it is nice but not affordable, I reply: We cannot afford to *not* do these things. We are already paying a high price for inequality, but it is just a down payment on what we will have to pay if we do not do something–and quickly. It is not just our economy that is at stake; we are risking our democracy.

As more of our citizens come to understand why the fruits of economic progress have been so unequally shared, there is a real danger that they will become open to a demagogue blaming the country's problems on others and making false promises of rectifying "a rigged system." We are already experiencing a foretaste of what might happen. It could get much worse.

About the Author

Joseph E. Stiglitz is a University Professor at Columbia University and chief economist at the Roosevelt Institute. He received the Nobel prize in economics in 2001. Stiglitz chaired President Bill Clinton's Council of Economic Advisers from 1995 to 1997 and served as the chief economist and senior vice president of the World Bank from 1997 to 2000. He chaired the Sarkozy commission (2008–2009) and an expert group (2013–2019) at the OECD for devising measures for well-being and sustainability.

Section 3: Technology

Algorithm Aids Search for Those Lost at Sea

By Scott Hershberger

When a craft sinks or goes missing at sea, search-and-rescue teams often rely on computer models to determine where to scour for survivors. Currently used models incorporate data from satellites and offshore sensors to predict a drifting object's path, producing maps of the areas where it is most likely to be found. If the initial search is unsuccessful, the models incorporate that information to update their predictions.

Now a team of researchers has developed a new algorithm to anticipate the location of drifting objects during the first three hours of a search. Instead of modeling the trajectory of a missing person, the tool identifies zones in the water called transient attracting profiles (TRAPs), where currents and waves conspire to pull in nearby objects. In the early hours of future search-and-rescue operations, factoring in these previously hidden "attractors" could prove crucial in saving lives.

The movement of ocean water can be represented mathematically as a velocity field, which in this case describes the speed and direction of water at each point on the surface. The new algorithm, described in May 2020 in *Nature Communications*, uses ocean wave data and forecast models of this field to find zones with the strongest pull. Though invisible in the water, each TRAP can be drawn on a map as a curve of about 100 to 1,000 meters long. As surface conditions change, TRAPs move slowly enough to drag objects along with them–similar to how a magnet moving under a table can pull an iron item along the tabletop.

Study lead author Mattia Serra of Harvard University and his colleagues field-tested the method by throwing GPS-tagged manikins in the turbulent waters just south of Martha's Vineyard in Massachusetts. Each manikin followed a different trajectory, but "they all clustered on the same TRAP," just as the algorithm predicted, Serra says. These initial tests were carried out close to shore, but the same mathematics predicts the presence of TRAPs in the open ocean.

Lawrence Stone, chief scientist at the scientific consulting company Metron and chief designer of the U.S. Coast Guard's standard search-and-rescue protocol, welcomes these "absolutely fascinating, impressive results." The study "fits right in with the sort of thing that we like to do, and it's good scientific research that holds up," says Stone, who was not involved in the new study. "Let's put it in."

Stone emphasizes that the TRAPs tool would supplement, not replace, existing models. A spokesperson confirms that the Coast Guard is working to incorporate TRAP curves into its prediction maps. As research continues, Serra and his colleagues hope to factor in the effects of wind and buoyancy, thus increasing the predictions' accuracy.

About the Author

Scott Hershberger was a 2020 AAAS Mass Media Fellow at Scientific American.

Color-Changing Fibers Unravel a Knotty Mystery

By Jeremy Hsu

How can mathematical theory predict the strongest knots in real life? The answer to that question has remained elusive despite humans relying on knots for thousands of years while fishing, hunting, building shelters and harnessing animal power. By using color-changing fibers to experimentally test a mathematical theory, researchers have begun developing more realistic models that can elucidate the rules governing stability in knots.

The mathematical theory of knots has typically focused on classifying their different entangled structures without accounting for mechanical stress and strain. To better predict knots' strength, Massachusetts Institute of Technology researchers experimented with those made from special photonic fibers that change color as they stretch. Doing so allowed the team to compare the mechanical-strain predictions of their mathematical models against those of the color-changing fiber experiments and therefore develop more sophisticated models that can simulate the intricacies of knots and, perhaps, more complex tangled structures.

"My collaborators developed beautiful fibers that change colors as you stretch them," says Jörn Dunkel, a mathematician at M.I.T. "They offered, for the first time, the ability to see in knots where the stress and strain goes in the fibers."

Such work allowed Dunkel and his colleagues to come up with three counting rules for predicting knot stability, as detailed in their study, published in the January 3, 2020, issue of the journal *Science*. They focused on analyzing the strength of different "bend" knots commonly used by sailors and climbers to tie two pieces of rope together.

The first rule suggests that knots gain stability when the two strands have a higher number of crossing points where they come in contact. The second one is that knots become more stable if strand segments at neighboring crossing points rotate in different directions and create opposing friction. And the third rule is that knots gain stability when strands slide tangentially against each other to create friction while being pulled in opposite directions.

Such findings have some practical implications. For example, the study suggests that a type of bend knot called the Zeppelin knot offers greater stability than the more popular alpine butterfly knot. "The nice thing is that with these basic rules, you can get quick intuition about which knots are more stable or less stable," Dunkel explains. "We can also use them as a starting point to explore more accurate models."

The M.I.T. team's work would not have been possible without the color-changing photonic fibers that were originally developed back in 2013. Such photonic fibers have a coating made from a periodic arrangement of two different types of elastic rubber, each with different material properties affecting their interaction with light. As a photonic fiber stretches under strain, its original color changes because the coating gets thinner and alters the original optical properties. Such fibers are examples of the same principle behind the rainbow of colors found in soap bubbles, butterflies and beetles.

"If, for instance, a relaxed fiber is red, it will change color through orange, yellow, green and blue to purple as it is stretched more and more," says Mathias Kolle, a mechanical engineer at M.I.T. and a co-author of the study. "In addition, we can tailor the spectral landscape of the fibers and play different tricks to generate other color progressions."

The experiments with the color-changing photonic fibers provided the reality check so that the mathematical models incorporating the same data could accurately predict the strongest types of knots. After simulating the strongest knots, researchers drew flat knot diagrams to examine the pattern of strands that make up each one. They also tested the real-world strength of knots by tying two pieces of Dyneema synthetic fiber together, hanging one end of the combined "rope" from a gripping mechanism and then attaching progressively heavier weights to the rope's dangling end.

The result is "a very interesting blend of experimental work and qualitative theoretical work," says Louis Kauffman, a professor emeritus in mathematics at the University of Illinois at Chicago, who was not involved in the M.I.T. study.

Promising though they may be, such methods are not without limitations. The flat-knot diagrams seem to do well in modeling real-world knot behavior so far, but more complex tangles may require 3-D modeling to fully account for how all the different forces are interacting. "This raises the very fascinating question of generalizing these models into full three-dimensional reality," Kauffman says.

Another limitation, Dunkel says, is that the mathematical models do not account for how different fiber materials affect the friction within a knot. Still, the team is eager to see if its models can find similar rules in systems beyond knots of fiber and string. For example, other research groups have discovered knot structures appearing within the liquid crystals used in technologies such as flat-screen TVs.

Separately, Kolle and his engineering team are continuing to develop their color-changing photonic fibers with the goal of making them more efficient in larger and more durable forms. So far, the researchers can make fibers about 15 centimeters in length in the lab, with each capable of thousands of stretching, color-changing cycles that traverse the entire visible spectrum.

The engineers have their eye on eventually making the fibers suitable for applications such as color-changing sportswear and fashion items, where alterations in color might help signal how well the fabric is holding up over time. Weaving such fibers into medical bandages might also help physicians and ordinary people visually check where bandages are applying maximum pressure to correctly staunch bleeding.

But commercialization possibilities aside, Kolle remains thrilled that the color-changing photonic fibers could help his colleagues untie a rather difficult knot of a problem. "It was very easy to get excited about it–the color-changing fibers being useful for really creative fundamental research," he says.

About the Author

Jeremy Hsu is a New York City–based writer who has contributed to publications such as Scientific American, IEEE Spectrum, Undark Magazine *and* Wired.

Math Can Help Build a Global Digital Community

By Cindy Lawrence

When the COVID-19 pandemic struck New York in March 2020 and the National Museum of Mathematics (MoMath) closed its doors to the public, little did we realize we were about to open a virtual door to the rest of the world.

A deadly public health crisis has taken over many aspects of our lives, affecting how we behave, what we read, what we talk about and what we fear. Yet within this new world context, we've discovered that, through the incredible richness of mathematics, MoMath is able to help create a global community of goodwill and mutual respect in an increasingly polarized world.

Math is a universal language that allows people to build new human connections even as many are continuing to physically isolate. At a time when divisiveness seems to have become all too common, MoMath is working to unite people of all ages and backgrounds through a shared enjoyment of mathematics.

Math is exceptionally well positioned to serve in this role of being a global, intergenerational online connector because at its heart, mathematics is an aesthetic pursuit. Despite what we may have learned from our past experiences, mathematics is not about learning *to do* arithmetic; it's about investigating the world around us and finding universal truths that can be both beautiful and inspiring. Math transcends language and culture, and it invites people of all ages and backgrounds to enter.

When New York City shut down, the museum launched into action, welcoming its first-ever online school field trip on the very same day that it hosted its last in-person visitors. Public programs continued to run thanks to the early presenters who bravely and enthusiastically jumped into a new world of digital communication.

Over the last year, MoMath has held more than 2,800 online programs that have attracted more than 87,000 participants from all 64 U.S. states and territories, as well as more than 100 countries, including Australia, Vietnam, Guatemala, Egypt, Argentina, Indonesia, China, Pakistan, Bhutan, Sweden, Chad, Zambia and France.

Because of the current global health crisis, people are prohibited from enjoying basic forms of human contact: firm handshakes conveying warmth; collaborative efforts while leaning over a computer screen or gathering around a whiteboard; engaging in stimulating conversation over a shared meal. While we all miss, even crave, these deeply human, everyday experiences and opportunities to interact in person, at MoMath we've discovered previously unseen virtual opportunities for sharing and learning.

In the physical world, it can be intimidating to ask a question, especially when those around you seem to be having no difficulty keeping up with the presentation. But now, we find that the cloak of anonymity that program attendees enjoy online actually encourages discourse. With participants identified only by a first name or perhaps even a pseudonym, inhibitions are removed. In our digital programs, we regularly see questions asked–and answered–by members of a global audience coming together as a group, disregarding geographic and cultural boundaries. That an online mathematics program can provide a supportive environment encouraging free inquiry and thoughtful discussion from the full diversity of humanity is humbling, inspiring and enlightening.

Digital math programming builds community, first and foremost, by providing opportunities for people from around the world to connect. MoMath has always strived to bring engaging math presenters to the stage, but audiences were limited to those who could make their way to Manhattan. Now, schoolchildren in India, teachers in Africa, families in Australia, college students in Europe and people in every state in our own country are able to come together to share experiences of mathematical wonder.

With the flip of a switch and the push of a button, those who have something they wish to share about math will find a global audience of enthusiastic people eager to hear from them. And those who might never have come into contact with renowned mathematicians suddenly have the opportunity, no matter where they are, to engage with some of the top minds in the field.

The online world of math also nurtures multigenerational connections. Grandparents who haven't seen their grandkids in months as a result of the pandemic find that math presents a fun and engaging way to entice the younger generation to join them in an hour of exploration and discovery. What a delight it was to see three generations of a single family come together at a family math event! And this intergenerational human connection occurs with total strangers as well: an elderly participant, self-identified as "alone and fallen on hard times," connected with a whip-smart young student at a recent evening gaming event, creating a truly magical moment of community.

MoMath online programming has created an ever-expanding universe of virtual learning, inviting all to enter, learn, enjoy and connect. When the pandemic gradually recedes and it is safe once again to welcome everyone back to the museum, we will keep our virtual doors propped wide open, allowing us to continue to unite people from around the world through the wonders of mathematics.

About the Author

Cindy Lawrence is executive director and CEO of the National Museum of Mathematics (MoMath), located in New York City.

The Mathematics of (Hacking) Passwords

By Jean-Paul Delahaye

At one time or another, we have all been frustrated by trying to set a password, only to have it rejected as too weak. We are also told to change our choices regularly. Obviously such measures add safety, but how exactly?

I will explain the mathematical rationale for some standard advice, including clarifying why six characters are not enough for a good password and why you should never use only lowercase letters. I will also explain how hackers can uncover passwords even when stolen data sets lack them.

ChOose#W!sely@*

Here is the logic behind setting hack-resistant passwords. When you are asked to create a password of a certain length and combination of elements, your choice will fit into the realm of all unique options that conform to that rule–into the "space" of possibilities. For example, if you were told to use six lowercase letters–such as, afzjxd, auntie, secret, wwwwww–the space would contain 26^6, or 308,915,776, possibilities. In other words, there are 26 possible choices for the first letter, 26 possible choices for the second, and so forth. These choices are independent: you do not have to use different letters, so the size of the password space is the product of the possibilities, or $26 \times 26 \times 26 \times 26 \times 26 \times 26 = 26^6$.

If you are told to select a 12-character password that can include uppercase and lowercase letters, the 10 digits and 10 symbols (say, !, @, #, $, %, ^, &, ?, / and +), you would have 72 possibilities for each of the 12 characters of the password. The size of the possibility space would then be 72^{12} (19,408,409,961,765,342,806,016, or close to 19×10^{21}).

That is more than 62 trillion times the size of the first space. A computer running through all the possibilities for your 12-character password one by one would take 62 trillion times longer. If your computer spent a second visiting the six-character space, it would have to devote two million years to examining each of the passwords in the 12-character space. The multitude of possibilities makes it impractical for a hacker to carry out a plan of attack that might have been feasible for the six-character space.

Calculating the size of these spaces by computer usually involves counting the number of binary digits in the number of possibilities. That number, N, is derived from this formula: 1 + integer($\log_2(N)$). In the formula, the value of $\log_2(N)$ is a real number with many decimal places, such as $\log_2(26^6) = 28.202638...$. The "integer" in the formula indicates that the decimal portion of that log value is omitted, rounding down to a whole number–as in integer(28.202638... 28). For the example of six lowercase letters above, the computation results in 29 bits; for the more complex, 12-character example, it is 75 bits. (Mathematicians refer to the possibility spaces as having entropy of 29 and 75 bits, respectively.) The French National Cybersecurity Agency (ANSSI) recommends spaces having a minimum of 100 bits when it comes to passwords or secret keys for encryption systems that absolutely must be secure. Encryption involves representing data in a way that ensures it cannot be retrieved unless a recipient has a secret code-breaking key. In fact, the agency recommends a possibility space of 128 bits to guarantee security for several years. It considers 64 bits to be very small (very weak); 64 to 80 bits to be small; and 80 to 100 bits to be medium (moderately strong).

Moore's law (which says that the computer-processing power available at a certain price doubles roughly every two years) explains why a relatively weak password will not suffice for long-term use: over time computers using brute force can find passwords faster. Although the pace of Moore's law appears to be decreasing, it is wise to take it into account for passwords that you hope will remain secure for a long time.

For a truly strong password as defined by ANSSI, you would need, say, a sequence of 16 characters, each taken from a set of 200 characters. This would make a 123-bit space, which would render the password close to impossible to memorize. Therefore, system designers are generally less demanding and accept low- or medium-strength passwords. They insist on long ones only when the passwords are automatically generated by the system, and users do not have to remember them.

There are other ways to guard against password cracking. The simplest is well known and used by credit cards: after three unsuccessful attempts, access is blocked. Alternative ideas have also been suggested, such as doubling the waiting time after each successive failed attempt but allowing the system to reset after a long period, such as 24 hours. These methods, however, are ineffective when an attacker is able to access the system without being detected or if the system cannot be configured to interrupt and disable failed attempts.

How Long Does It Take to Search All Possible Passwords?

For a password to be difficult to crack, it should be chosen randomly from a large set, or "space," of possibilities. The size, ***T***, of the possibility space is based on the length, ***A***, of the list of valid characters in the password and the number of characters, ***N***, in the password. The size of this space ($T = A^N$) may vary considerably.

Each of the following examples specifies values of ***A***, ***N***, ***T*** and the number of hours, ***D***, that hackers would have to spend to try every permutation of characters one by one. ***X*** is the number of years that will have to pass before the space can be checked in less than one hour, assuming that Moore's law (the doubling of computing capacity every two years) remains valid. I also assume that in 2019, a computer can explore a billion possibilities per second. I represent this set of assumptions with the following three relationships and consider five possibilities based on values of ***A*** and ***N***:

RELATIONSHIPS

$T = A^N$

$D = T/(10^9 \times 3{,}600)$

$X = 2 \log_2[T/(10^9 \times 3{,}600)]$

RESULTS

If $A = 26$ and $N = 6$, then $T = 308{,}915{,}776$
$D = 0.0000858$ computing hour
$X = 0$; it is already possible to crack all passwords in the space in under an hour

---

If $A = 26$ and $N = 12$, then $T = 9.5 \times 10^{16}$
$D = 26{,}508$ computing hours
$X = 29$ years before passwords can be cracked in under an hour

---

If $A = 100$ and $N = 10$, then $T = 10^{20}$
$D = 27{,}777{,}777$ computing hours
$X = 49$ years before passwords can be cracked in under an hour

---

If $A = 100$ and $N = 15$, then $T = 10^{30}$
$D = 2.7 \times 10^{17}$ computing hours
$X = 115$ years before passwords can be cracked in under an hour

---

If $A = 200$ and $N = 20$, then $T = 1.05 \times 10^{46}$
$D = 2.7 \times 10^{33}$ computing hours
$X = 222$ years before passwords can be cracked in under an hour

---

Weaponizing Dictionaries and Other Hacker Tricks

Quite often an attacker succeeds in obtaining encrypted passwords or password "fingerprints" (which I will discuss more fully later) from a system. If the hack has not been detected, the interloper may have days or even weeks to attempt to derive the actual passwords.

To understand the subtle processes exploited in such cases, take another look at the possibility space. When I spoke earlier of bit size and password space (or entropy), I implicitly assumed that the user consistently chooses passwords at random. But typically the choice is not random: people tend to select a password they can remember (locomotive) rather than an arbitrary string of characters (xdichqewax).

This practice poses a serious problem for security because it makes passwords vulnerable to so-called dictionary attacks. Lists of commonly used passwords have been collected and classified according to how frequently they are used. Attackers attempt to crack passwords by going through these lists systematically. This method works remarkably well because, in the absence of specific constraints, people naturally choose simple words, surnames, first names and short sentences, which considerably limits the possibilities. In other words, the nonrandom selection of passwords essentially reduces possibility space, which decreases the average number of attempts needed to uncover a password.

Below are the first 25 entries in one of these password dictionaries, listed in order, starting with the most common one. (I took the examples from a database of five million passwords that was leaked in 2017 and analyzed by SplashData.)

1. 123456
2. password
3. 12345678
4. qwerty
5. 12345
6. 123456789
7. letmein
8. 1234567
9. football
10. iloveyou
11. admin
12. welcome
13. monkey
14. login

15. abc123
16. starwars
17. 123123
18. dragon
19. passw0rd
20. master
21. hello
22. freedom
23. whatever
24. qazwsx
25. trustno1

If you use password or iloveyou, you are not as clever as you thought! Of course, lists differ according to the country where they are collected and the websites involved; they also vary over time.

For four-digit passwords (for example, the PIN code of SIM cards on smartphones), the results are even less imaginative. In 2013, based on a collection of 3.4 million passwords each containing four digits, the DataGenetics website reported that the most commonly used four-digit sequence (representing 11 percent of choices) was 1234, followed by 1111 (6 percent) and 0000 (2 percent). The least-used four-digit password was 8068. Careful, though, this ranking may no longer be true now that the result has been published. The 8068 choice appeared only 25 times among the 3.4-million four-digit sequences in the database, which is much less than the 340 uses that would have occurred if each four-digit combination had been used with the same frequency. The first 20 series of four digits are: 1234; 1111; 0000; 1212; 7777; 1004; 2000; 4444; 2222; 6969; 9999; 3333; 5555; 6666; 1122; 1313; 8888; 4321; 2001; 1010.

Even without a password dictionary, using differences in frequency of letter use (or double letters) in a language makes it possible to plan an effective attack. Some attack methods also take into account that, to facilitate memorization, people may choose passwords that have a certain structure—such as A1=B2=C3, AwX2AwX2 or O0o.lli. (which I used for a long time)—or that are derived by combining several simple strings, such as password123 or johnABC0000. Exploiting such regularities makes it possible for hackers to speed up detection.

Making Hash of Hackers

As the main text explains, instead of storing clients' passwords, Internet servers store the "fingerprints" of these passwords: sequences of characters that are derived from the passwords. In the event of an attack, the use of fingerprints can make it is very difficult, if not impossible, for hackers to use what they find.

The transformation is achieved by using algorithms known as cryptographic hash functions. These are meticulously developed processes that transform a data file, F, however long it may be, into a sequence, *h*(F), called a fingerprint of F. For example, the hash function SHA256 transforms the phrase "Nice weather" into:

DB0436DB78280F3B45C2E09654522197D59EC98E7 E64AEB967A2A19EF7C394A3

(64 hexadecimal, or base 16, characters, which is equivalent to 256 bits).

Changing a single character in the file completely alters its fingerprint. For example, if the first character of Nice weather is changed to lowercase (nice weather), the hash SHA256 will generate another fingerprint:

02C532E7418CD1B57961A1B090DB6EC37B3C58380AC0 E6877F3B6155C974647E

Good hash functions produce fingerprints that are similar to those that would be obtained if the fingerprint sequence was uniformly chosen at random. In particular, for any possible random result (a sequence of 64 hexadecimal characters), it is impossible to find a data file F with this fingerprint in a reasonable amount of time.

There have been several generations of hash functions. The SHA0 and SHA1 generations are obsolete and are not recommended. The SHA2 generation, including SHA256, is considered secure.

The Take-Home for Consumers

Taking all this into account, properly designed websites analyze the passwords proposed at the time of their creation and reject those that would be too easy to recover. It is irritating, but it's for your own good.

The obvious conclusion for users is that they must choose their passwords randomly. Some software does provide a random password. Be aware, however, that such password-generating software may, deliberately or not, use a poor pseudo-random generator, in which case what it provides may be imperfect.

You can check whether any of your passwords has already been hacked by using a web tool called Pwned Passwords. Its database includes more than 500 million passwords obtained after various attacks.

I tried e=mc2e=mc2, which I liked and believed to be secure, and received an unsettling response: "This password has been seen 114 times before." Additional attempts show that it is difficult to come up with easy-to-memorize passwords that the database does not know. For example, aaaaaa appeared 395,299 times; a1b2c3d4, 113,550 times; abcdcba, 378 times; abczyx, 186 times; acegi, 117 times; clinton, 18,869 times; bush, 3,291 times; obama, 2,391 times; trump, 859 times.

It is still possible to be original. The website did not recognize the following six passwords, for example: eyahaled (my name spelled backward); bizzzzard; meaudepace and modeuxpass (two puns on the French for "password"); abcdef2019; passwaurde. Now that I've tried them, I wonder if the database will add them when it next updates. In that case, I won't use them.

Advice for Websites

Websites, too, follow various rules of thumb. The National Institute of Standards and Technology recently published a notice recommending the use of dictionaries to filter users' password choices.

Among the rules that a good web server designer absolutely must adhere to is to not store plaintext lists of usernames and passwords on the computer used to operate the website.

The reason is obvious: hackers could access the computer containing this list, either because the site is poorly protected or because the system or processor contains a serious flaw unknown to anyone except the attackers (a so-called zero-day flaw), who can exploit it.

One alternative is to encrypt the passwords on the server: use a secret code that transforms them via an encryption key into what will appear to be random character sequences to anyone who does not possess the decryption key. This method works, but it has two disadvantages. First, it requires decrypting the stored password every time to compare it with the user's entry, which is inconvenient. Second, and more seriously, the decryption necessary for this comparison requires storing the decryption key in the website computer's memory. This key may therefore be detected by an attacker, which brings us back to the original problem.

A better way to store passwords is through what are called hash functions that produce "fingerprints." For any data in a file–symbolized as F–a hash function generates a fingerprint. (The process is also called condensing or hashing.) The fingerprint–*h*(F)–is a fairly short word associated with F but produced in such a way that, in practice, it is impossible to deduce F from *h*(F). Hash functions are said to be one-way: getting from F to *h*(F) is easy; getting from *h*(F) to F is practically impossible. In addition, the hash functions used have the characteristic that even if it is possible for two data inputs, F and F', to have the same fingerprint (known as a collision), in practice for a given F, it is almost impossible to find an F' with a fingerprint identical to F.

Using such hash functions allows passwords to be securely stored on a computer. Instead of storing the list of paired usernames and passwords, the server stores only the list of username/fingerprint pairs.

When a user wishes to connect, the server will read the individual's password, compute the fingerprint and determine whether it corresponds to the list of stored username/fingerprint pairs associated with that username. That maneuver frustrates hackers because even if they have managed to access the list, they will be unable to derive the users' passwords, inasmuch as it is practically impossible to go from fingerprint to password. Nor can they generate another password with an identical fingerprint to fool the server because it is practically impossible to create collisions.

Still, no approach is foolproof, as is highlighted by frequent reports of the hacking of major sites. In 2016, for example, data from a billion accounts were stolen from Yahoo!

For added safety, a method known as salting is sometimes used to further impede hackers from exploiting stolen lists of username/fingerprint pairs. Salting is the addition of a unique random string of characters to each password. It ensures that even if two users employ the same password, the stored fingerprints will differ. The list on the server will contain three components for each user: username, fingerprint derived after salt was added to the password, and the salt itself. When the server checks the password entered by a user, it adds the salt, computes the fingerprint and compares the result with its database.

Even when user passwords are weak, this method considerably complicates the hacker's work. Without salting, a hacker can compute all the fingerprints in a dictionary and see those in the stolen data; all the passwords in the hacker's dictionary can be identified. With salting, for every salt used, the hacker must compute the salted fingerprints of all the passwords in the hacker's dictionary. For a set of 1,000 users, this multiplies by 1,000 the computations required to use the hacker's dictionary.

Survival of the Fittest

It goes without saying that hackers have their own ways of fighting back. They face a dilemma, though: their simplest options either take a lot of computing power or a lot of memory. Often neither option is viable. There is, however, a compromise approach known as the rainbow table method (see "Rainbow Tables Help Hackers").

In the age of the Internet, supercomputers and computer networks, the science of password setting and cracking continues to evolve—as does the relentless struggle between those who strive to protect passwords and those who are determined to steal, and potentially abuse, them.

Rainbow Tables Help Hackers

Say you are a hacker looking to exploit data that you have acquired. These data consist of username/fingerprint pairs, and you know the hash function (see "Making Hash of Hackers"). The password is contained in the possibility space of strings of 12 lowercase letters, which corresponds to 56 bits of information and 26^{12} (9.54×10^{16}) possible passwords.

At least two strong approaches are open to you:

Method 1. You scroll through the entire space of passwords. You calculate the fingerprint, $h(P)$, for each password, checking to see whether it appears in the stolen data. You do not need a lot of memory, because prior results are deleted with each new attempt, although you do, of course, have to keep track of the possibilities that have been tested.

Scrolling through all the possible passwords in this way takes a long time. If your computer runs a billion tests per second, you will need $26^{12}/(10^9 \times 3{,}600 \times 24)$ days (1,104 days), or about three years to complete the task. The feat is not impossible; if you happen to have a computer network of 1,000 machines, one day will suffice. It is not feasible, however, to repeat such a calculation every time you wish to test additional data, such as if you obtain a new set of username/fingerprint pairs. (Because you have not saved the results of your computations, you would need an additional 1,104 days to process the new information.)

Method 2. You say to yourself, "I'll compute the fingerprints of all possible passwords, which will take time, and I'll store the resulting fingerprints in a big table. Then I'll have to find only a password fingerprint in the table to identify the corresponding password in the stolen data."

You will need $(9.54 \times 10^{16}) \times (12 + 32)$ bytes of memory because the task requires 12 bytes for the password and 32 bytes for the fingerprint if the fingerprint contains 256 bits (assuming an SHA256 function). That's 4.2×10^{18} bytes, or 4.2 million hard disks with a capacity of one terabyte.

This memory requirement is too large. Method 2 is no more feasible than method 1. Method 1 requires too many computations, and method 2 requires too much memory. Both cases are problematic: either each new password takes too long to compute, or precomputing all possibilities and storing all the results is too large a task.

Is there some compromise that requires less computing power than method 1 and less memory than required for method 2? Indeed, there is. In 1980 Martin Hellman of Stanford University suggested an approach that was improved in 2003 by Philippe Oechslin of the Swiss Federal Institute of Technology in Lausanne and further refined more recently by Gildas Avoine of the National Institute of Applied Sciences of Rennes (INSA Rennes) in France. It demands less computing power than method 1 in exchange for using a little more memory.

The Beauty of the Rainbow

Here is how it works: First, we need a function *R* that transforms a fingerprint *h(P)* into a new password *R(h(P))*. One might, for instance, consider fingerprints as numbers written in the binary numeral system and consider passwords as numbers written in the K numeral system, where K is the number of allowable symbols for passwords. Then the function *R* converts data from the binary numeral system to the K numeral system. For every fingerprint *h(P)*, it computes a new password *R(h(P))*.

Now, with this function *R*, we can precompute data tables called rainbow tables (so named perhaps because of the multicolored way these tables are depicted).

To generate a data point in this table, we start from a possible password P_0, compute its fingerprint, $h(P_0)$ and then compute a new possible password *R(h*(PO)), which becomes P1. Next, we continue this process from P1. Without storing anything other than P_0, we compute the sequence P_1, P_2,... until the fingerprint starts with 20 zeros; that fingerprint is designated $h(P_n)$. Such a fingerprint occurs only once in about 1,000,000 fingerprints because the result of a hash function is similar to result of a uniform random draw, and 220 is roughly equal to 1,000,000. The password/fingerprint pair [P_0, $h(P_n)$], containing the fingerprint that starts with 20 zeros is then stored in the table.

A very large number of pairs of this type are computed. Each password/fingerprint pair [P_0, $h(P_n)$] represents the sequence of passwords P_0, P_1,... P_n and their fingerprints, but the table does not store those intermediate calculations. The table thus lists many password/fingerprint pairs and represents many more (the intermediates, such as P_1 and P_2, that can be derived from the listed pairs). But, of course, there may be gaps: some passwords may be absent from all the chains of calculations.

For a good database with almost no gaps, the memory needed to store the calculated pairs is a million times smaller than that needed for method 2, as described earlier. That is less than four one-terabyte hard disks. Easy. Also, as will be seen, using the table to derive passwords from stolen fingerprints is quite doable.

Let us see how the data stored on the hard disks makes it possible to determine a password in a given space in just a few seconds. Assume that there are no gaps; precomputation of the table takes into account all the passwords of a designated type—for example, 12-character passwords taken from the 26 letters of the alphabet.

A fingerprint f_0 in a stolen data set can be used to reveal the associated password in the following way. Calculate $h(R(f_0))$ to arrive at a new fingerprint, f_1, then calculate $h(R(f_1))$ to get f_2, and so on, until you get to a fingerprint that begins with 20 zeros: f_m. Then check the table to see which original password, P_0, the fingerprint f_m is associated with. Based on P_0, calculate the passwords and fingerprints *h1, h2,...* that follow until you inevitably generate the original fingerprint f_0, designated h_k. The password you are looking for is the one that gave rise to h_k—in other words, $R(h_{k-1})$, which is one step earlier in the chain of calculations.

The computation time required is what it takes to look for f_m in the table plus the time needed to compute the sequence of fingerprints from the associated password (h_1, h_2,..., h_k)—which is about a million times shorter than the time needed to compute the table itself. In other words, the time needed is quite reasonable.

Thus, doing a (very long) precomputation and storing only part of the results makes it possible to retrieve any password with a known fingerprint in a reasonable amount of time.

To summarize, by knowing the beginning and end of each chain of computations (the only things that are stored during precomputation), a hacker can retrieve any password from a fingerprint. In somewhat simplistic terms, starting from a stolen fingerprint—call it fingerprint X—a hacker would apply the R and *h* functions repeatedly, calculating a series of passwords and fingerprints until reaching a fingerprint with 20 zeros in front of it. The hacker would then look up that final fingerprint in the table (Fingerprint C in the example below) and identify its corresponding password (Password C).

Sample Table Excerpt
Password A—Fingerprint A
Password B—Fingerprint B
Password C–Fingerprint C
Password D—fingerprint D

Next, the hacker would apply the *h* and *R* functions again, beginning with the identified password, continuing on until one of the resulting fingerprints in the chain matches the stolen fingerprint:

Sample Calculation
Password C --> fingerprint 1 --> password 2-- --> fingerprint 2 --> password 3.... --> password 22-- -->**fingerprint 23** [a match to fingerprint X!]

The match (fingerprint 23) would indicate that the previous password (password 22), from which the fingerprint was derived, is the one linked to the stolen fingerprint.

Many computations must be done to establish the first and last column of the rainbow table. By storing only the data in these two columns and by recomputing the chain, hackers can identify any password from its fingerprint.

About the Author

Jean-Paul Delahaye is a professor emeritus of computer science at the University of Lille in France and a researcher at the Research Center in Computer Science, Signal and Automatics of Lille (CRIStAL). He recently published Les Mathématiciens Se Plient au Jeu *(Belin, 2017), a French collection of articles from* Pour la Science.

Why History Should Matter to Code Breakers

By Craig P. Bauer

A science librarian once told me that all the literature on biology, chemistry and other sciences more than five years old could be thrown out and it wouldn't make any difference. She said it was only mathematicians and some physicists who cared about anything older. With science and technology advancing so rapidly that anything old enough to have a little dust on it is obsolete, does history really matter?

In my field, cryptology–the study of codes and ciphers–it definitely does.

Whereas encryption algorithms are continually being improved, upgrades are not always automatic. There are plenty of old systems still in use. And when it comes to ciphers used by bad guys of all sorts, old can mean *centuries*. Some criminals believe the National Security Agency can break all of the current systems so they fall back on older, weaker ciphers. This is ridiculous for a number of reasons–not the least of which, it is the FBI, not the NSA, that American citizens engaged in criminal activities need to worry about.

In any case, the FBI's Cryptanalysis and Racketeering Records Unit (CRRU) regularly sees Vigenère ciphers, for example, despite the fact this method goes back to the 16th century and a method for deciphering it was published in 1863. Of course, criminals make all sorts of mistakes, and failing to learn from history is among them.

Criminals don't just fall back on older ciphers out of paranoia; in some cases they don't have access to the technology needed to keep their encryption up to date. A great example of this is law-breakers serving time. About 80 percent of the CRRU's workload consists of prison ciphers. Imprisoned gang members are often still running operations and giving orders. Their resources are limited, but anyone with pencil and paper can write in cipher. The Aryan Brotherhood is the source of endless FBI headaches in this regard. One doesn't need to belong to a gang to hide information with codes or ciphers, however. There are numerous serial killers who carried out their crimes without accomplices but nonetheless wrote in cipher, sometimes to newspapers and sometimes in their personal journals. Examples include Zodiac, a Zodiac copycat killer in 1990s New York City, the Unabomber, the mass murderer known as BTK (Bind, Torture, Kill) and more. Old paper-and-pencil-type ciphers are still important.

It's common for a person enciphering a message to be overconfident as to the level of protection thus acquired. This is nothing new. During World War II some Nazis realized the Enigma cipher machines on which they relied so heavily could theoretically be cracked, but they didn't think anyone would make the intense effort required to actually do it. The fact mathematicians from Poland, England and the U.S. did is now well known–among other things, it was the plot of the 2014 movie *The Imitation Game*–but the story was kept secret until the 1970s. The people who were in the know (more than 10,000 of them) kept their mouths shut for decades. Meanwhile various versions of the Enigma machine and other similar devices continued to be used–in some cases on into the 1990s. Once again, "historical" ciphers stayed relevant.

In the digital world and the throwaway society we've become, where only the latest and greatest devices and technology will suffice, it can be easy to think of everything in the same way. But codes and ciphers have a long track record of being recycled, and professionals must realize that what's old is new again.

Yes, technology is always changing. But there are hundreds of examples of centuries-old ciphers that still remain unsolved to this day. They illustrate that newer is not always better as well as the importance of thinking holistically when approaching the mysteries that remain before us.

About the Author

Craig P. Bauer is a professor of mathematics at York College of Pennsylvania and author of Unsolved! The History and Mystery of the World's Greatest Ciphers from Ancient Egypt to Online Secret Societies.

Section 4: Health

Small Numbers Can Have Huge Impacts on Climate and Health

By Naomi Oreskes

Quantifying things is an overarching goal in science, but recent events have left me pondering just why that is. After all, we've long known that numbers can be precise but inaccurate. The textbook exemplar is Lord Kelvin, the 19th-century British physicist who insisted that unless you could quantify a thing, your knowledge of it was of a "meager and unsatisfactory kind."

Yet he was responsible for one of history's most infamous examples of a fallacious quantitative argument. It concerned the geologic principle of uniformitarianism, which states that processes we can observe can be used to interpret Earth's history. Kelvin insisted it was wrong because it had led geologists to conclude that Earth was billions of years old. Kelvin's own calculations ended up placing the age at no more than 20 million to 40 million years; ergo the methods geologists had used must be faulty.

Kelvin's paper on the topic is perhaps the most arrogant in the history of science: the body of *The "Doctrine of Uniformity" in Geology Briefly Refuted* is one paragraph long, with one additional paragraph laying out his calculations. But it was Kelvin who ended up being refuted. He had made (at least) two faulty assumptions: one, that Earth began in a molten state and has been cooling ever since, and two, that there is no source of heat other than that left over from the planet's formation. (Today scientists think our planet began as gas and dust; we know that radioactive decay is a major source of heat and that Earth is 4.5 billion years old.)

Calculations can never be better than the assumptions that guide them, but if numbers are fallible, then what exactly is their point? One answer is that they give us the ability to compare one thing against another using a common scale. It's a whole lot easier to compare 1,000 to 100 than it is to compare "canoe" to "pineapple."

Another is specificity: we know how much larger 1,000 is than 100, but it's not necessarily clear what we mean when we call something simply "large" or "small." Indeed, two recent scientific debates both hinged on the question "How small is small"?

The first is about the risks of eating red meat. Last year a group of researchers argued that Americans should not bother to reduce their consumption of red and processed meats, because any health benefits would be "small." The second is about the impact of fossil-fuel subsidies. A paper in *Nature* examined whether eliminating them would have much of an effect on carbon dioxide emissions. The authors concluded that the effect would be "limited." (Full disclosure: I am an author on a paper challenging that finding.)

How are we to judge these claims? Critics have a point when they suggest that we should not expect people to change their lives–or governments to change their policies–if the benefits of those changes have not been demonstrated or if they have been demonstrated but are insubstantial. But how do we judge what is substantial and what is not? In the case of red meat, the authors wrote: "Dose-response meta-analysis results from 17 cohorts with 2.2 million participants provided low-certainty evidence that decreasing unprocessed red meat intake may result in a very small reduction of overall lifetime cancer mortality (7 fewer events per 1,000 persons with a decrease of 3 servings/wk)."

Is seven fewer cancer deaths per 1,000 people "very small"? In a population the size of the U.S.'s (about 330 million), it would mean 2.3 million people! That lifetime total amounts to an average of roughly 30,000 deaths a year. In comparison, as of late March this year, the Centers for Disease Control and Prevention estimated flu deaths in the 2019–2020 season (which are calculated separately from those caused by COVID-19) at 23,000 to 59,000. It would be accurate–and perhaps more informative–to say that eating three fewer servings of red meat a week could save about as many lives as eliminating deaths from seasonal flu.*

In the case of fossil fuels, the authors wrote: "Subsidy removal would lead to a small decrease in global CO2 emissions:

0.5–2 gigatons ... This is much less than the Nationally Determined Contributions (NDCs) from the Paris Agreement, which add up to a decrease of between 4–8 Gt from fossil fuels and industry." Yes, a range of 0.5 to two is less than four to eight, but is it "much less"? Maybe, but the authors would have been equally correct if they had said that subsidy removal would amount to at least 6 percent of the Paris commitments and possibly as much as 50 percent.

Admittedly, scientists are often admonished to speak in plain English, and "small" and "very small" are plain English. But when the numbers they refer to are *not* small, they become misleading.

*__Editor's Note (6/24/20)__: This paragraph has been edited after posting to clarify the comparison between deaths associated with red meat and seasonal influenza.

About the Author

Naomi Oreskes is a professor of the history of science at Harvard University. She is author of Why Trust Science? *(Princeton University Press, 2019) and co-author of* Discerning Experts *(University of Chicago, 2019).*

The Heisenberg Uncertainty Principle of Social Science Modeling

By Ben Klemens

Arguably, all of scientific inquiry in modern times begins with some sort of model. A model takes different parameters you are studying and uses them to make some claim about how our world works. It is a reduction of reality with the aim of reconstructing a picture of truth, whether about the spread of disease, the population of a toad species or the number of people who will move in 2030.

But as the number of things we're trying to study grows, the chance of getting even close to the target reality falls. The reason for this trade-off is the "curse of dimensionality." It is not a rule of thumb or a limit due to measurement errors, but as much a mathematical fact as the Pythagorean theorem—and it puts fundamental limits on what economics and other social sciences can describe. The curse of dimensionality is why our estimates of how a disease will behave will always have imprecision.

Dimensions most commonly refers to the space and time we occupy, but it can also mean any set of measurable things that are independent of each other. For example, let's say we want a model of how a public health campaign would affect COVID-19 spread. We might use factors such as the estimated incubation period of the disease under given conditions (call it X), the percentage of people who would wear masks in public given a public health campaign (Y), an estimate of person-to-person transmission likelihood (Z), and so on, to estimate patterns of spread. To make predictions about the effects of our ad campaign, we will need to find numeric values for X, Y and Z (i.e., "Fit the model to data").

The model has three independent dimensions, so the model parameters X, Y and Z can be read as points in 3-D space. If we get our best fit for what X, Y and Z should be using our real-world data and modeling techniques, will our estimate come close to the true value (which we would only be able to observe directly if we were omniscient)?

To answer this question, we need to think about how shapes behave in different dimensions.

If you have some solid shape with a thin shell around it, the shell holds a surprising amount of the volume. Get an orange from the supermarket nine centimeters in diameter, whose peel is just 0.45 cm thick. About 25 percent of the orange's volume is in the peel.

What if your orange came in a gift box, exactly big enough that the fruit touches all sides? Stepping back to two dimensions for a moment, a circle takes up 78.5 percent of the volume of its tightest-fitting square. In three dimensions the orange in its box 47.6 percent of the volume of the box and the rest is empty air. As the number of dimensions rise, the percent of inside-the-box volume that is the fruit itself shrinks further. A four-dimensional ball is 30.8 percent of the volume of the box. By nine dimensions, the tightest-fitting box is 99.54 percent empty. Or if you are an optimist, the box is 0.46 percent full.

Now, let's think about the COVID-19 model as if it existed as a shape in three-dimensional space. Imagine the center of the box to be the true value of X, Y and Z, and the tight-fitting box to be the range of our best guesses regarding each parameter by itself. Define "close" as being inside the sphere in the center of the shell or tight-fitting box. The word "close" has the obvious physical interpretation, but also makes sense in information space, where we need our estimate of X, Y and Z to be within a short distance to the true value. The fact that a point chosen haphazardly in the higher-dimensional box has such a small chance of being close to the center is an example of the curse of dimensionality.

Say we want our model to be more descriptive. A public health campaign could cause people go to the supermarket A percent less often, and induce B percent to work from home, and C percent to stop taking public transportation. Adding these parameters brings us to a six dimensional model (A, B, C, X, Y, Z), and we could easily brainstorm three or four more. If we could put good bounds on the numeric range of each variable, we could put our estimate in a tight-fitting box around the true nine-dimensional parameter value–which gives us a 0.46 percent chance of our full model with all nine moving parts being close to truth.

This is the balancing act of model design. We want to make our models more descriptive by adding more interacting elements, but the curse of dimensionality all but guarantees that if you try to fit a model with a large number of parameters to data, your fit won't be close. We can get good estimates of the effect of a public health campaign in a broad context with few details, or we can get an imprecise estimate in a focused and detailed setting, but a high level of detail and a precise estimate for all those parameters is near impossible.

The solutions for a researcher are to avoid simultaneously estimating parameter sets, to accept models with limited scope and few moving parts, to structure models with many more assumptions to reduce informational dimensions, or to put extraordinary work into pegging down every parameter with great precision. In short, resist the desire to fit the latest data set to a model of everything. The solution for readers of research is to accept the limitations of models that do not try to be a theory of everything, and maintain skepticism of models that seem to flout the curse.

About the Author

Ben Klemens is a computational social scientist living in Washington, DC, and the co-host of the podcast "Pod Paper Scissors." He holds a Ph.D. in social sciences from Caltech. His publications include papers on the characteristics of within-U.S. movers, theoretical modeling of patent valuation, and measuring poverty from space.

This Formula Calculates How Many Calories You Burn If You're Doing Absolutely Nothing

By Florian Freistetter

There's a plate of cookies on my desk. I put it there myself, so I can only blame myself for avoiding work with the prospect of a sweet snack. I don't even want to know exactly how many calories there are in a single little cookie. Yet this is exactly what you should know if you want to lose weight.

The calorie is a unit of energy and is actually out-of-date. The official measure should be given in joules. But especially when it comes to energy that we add to our body by ingesting food, we still like to use the calorie. The same applies to mathematical formulas such as the basal metabolic rate, the amount of energy the body uses at rest for basic functions, including breathing and staying warm:

$$G = \left(\frac{10m}{1kg} + \frac{6.25h}{1cm} - \frac{5a}{1yr} + s \right)$$

If you insert the weight m (in kilograms), the height h (in centimeters) and the age a (in years) and choose the factor s according to gender (+5 for men and −161 for women), you get the Mifflin–St. Jeor formula that shows the basal metabolic rate in kilocalories per day. This is the amount of energy that our body needs to keep up its basic functions.

The value from the formula is not exact. In reality, the basal metabolic rate depends on many other factors than just weight, height and age. Everyone has a different metabolism, and hardly anyone lies around motionless or is solely concerned with maintaining basic bodily functions. Any kind of activity increases energy expenditure.

Only in certain cases of bedridden sick people does the basal metabolic rate approximate the actual energy requirement.

Nevertheless, it is good to know just how much energy you can ingest if you want to lose weight. To do so, the basic rule is that you have to use more energy than you consume. And if you want to avoid the yo-yo effect–endless cycles of losing weight loss and then gaining it all back–you should make sure that the change in the energy balance is reasonably sustainable.

The many different diets in various magazines are generally of little help, even if they try to impress you with the math, as in the case of the "minus 500" formula. If you believe the puffery, then it is considered to be an ingenious formula. It is thought to be an equation developed by nutrition experts, a "mathematically proved" strategy to lose weight without all the problems that one usually has with diets.

Minus 500 turns up frequently in weight-loss programs. The formula is simple: multiply body weight in kilograms by 30 and subtract 500, and the result of the calculation is the amount of kilocalories that you can consume per day. And as much as I like to rant about it, this formula actually works. That is because body weight times 30 is a simple estimate for the basal metabolic rate. It is, of course, much less precise than the Mifflin–St. Jeor formula, but it is not drawn completely out of thin air. And if you subtract 500 kilocalories from your basal metabolic rate and limit your daily energy intake to this value, you will definitely consume less energy than you use.

From a purely mathematical point of view, there is nothing wrong with the minus 500 formula. But the question of whether it is always appropriate to reduce your energy intake in this simple way should be checked with a medical professional beforehand. In the end, however, I have to grumble a little. The math is merely giving weight to a simple fact. In the end, the "ingenious formula" developed by the "experts" only comes down to what everyone actually knows: if you want to lose weight, you have to eat less and exercise more—which is why my cookie and I are going to take a walk.

About the Author

Florian Freistetter is an astronomer, author and science cabaret artist with Science Busters, a group that blends comedy and science. Freistetter's Formula World is his regular online column on Spektrum.de, *the online science portal for* Spektrum der Wissenschaft, Scientific American's *German-language sister publication.*

When Assessing Novel Risks, Facts Are Not Enough

By Baruch Fischhoff

Psychologists study how humans make decisions by giving people "toy" problems. In one study, for example, my colleagues and I described to subjects a hypothetical disease with two strains. Then we asked, "Which would you rather have? A vaccine that completely protects you against one strain or a vaccine that gives you 50 percent protection against both strains?" Most people chose the first vaccine. We inferred that they were swayed by the phrase about complete protection, even though both shots gave the same overall chance of getting sick.

But we live in a world with real problems, not just toy ones–situations that sometimes require people to make life-and-death decisions in the face of incomplete or uncertain knowledge. Years ago, after I had begun to investigate decision-making with my colleagues Paul Slovic and the late Sarah Lichtenstein, both at the firm Decision Research in Eugene, Ore., we started getting calls about non-toy issues–calls from leaders in industries that produced nuclear power or genetically modified organisms (GMOs). The gist was: "We've got a wonderful technology, but people don't like it. Even worse, they don't like us. Some even think that we're evil. You're psychologists. Do something."

We did, although it probably wasn't what these company officials wanted. Instead of trying to change people's minds, we set about learning how they *really* thought about these technologies. To that end, we asked them questions designed to reveal how they assessed risks. The answers helped us understand why people form beliefs about divisive issues such as nuclear energy–and today, climate change–when they do not have all the facts.

Intimations of Mortality

To start off, we wanted to figure out how well the general public understands the risks they face in everyday life. We asked groups of laypeople to estimate the annual death toll from causes such as drowning, emphysema and homicide and then compared their estimates with scientific ones. Based on previous research, we expected that people would make generally accurate predictions but that they would overestimate deaths from causes that get splashy or frequent headlines–murders, tornadoes–and underestimate deaths from "quiet killers," such as stroke and asthma, that do not make big news as often.

Overall, our predictions fared well. People overestimated highly reported causes of death and underestimated ones that received less attention. Images of terror attacks, for example, might explain why people who watch more television news worry more about terrorism than individuals who rarely watch. But one puzzling result emerged when we probed these beliefs. People who were strongly opposed to nuclear power believed that it had a very low annual death toll. Why, then, would they be against it? The apparent paradox made us wonder if by asking them to predict average annual death tolls, we had defined risk too narrowly. So, in a new set of questions we asked what risk really meant to people. When we did, we found that those opposed to nuclear power thought the technology had a greater potential to cause widespread catastrophes. That pattern held true for other technologies as well.

To find out whether knowing more about a technology changed this pattern, we asked technical experts the same questions. The experts generally agreed with laypeople about nuclear power's death toll for a typical year: low. But when they defined risk themselves, on a broader time frame, they saw less potential for problems. The general public, unlike the experts, emphasized what could happen in a very bad year. The public and the experts were talking past each other and focusing on different parts of reality.

Understanding Risk

Did experts always have an accurate understanding of the probabilities for disaster? Experts analyze risks by breaking complex problems into more knowable parts. With nuclear power, the parts might include the performance of valves, control panels, evacuation schemes and cybersecurity defenses. With GMO crops, the parts might include effects on human health, soil chemistry and insect species.

The quality and accuracy of a risk analysis depend on the strength of the science used to assess each part. Science is fairly strong for nuclear power and GMOs. For new technologies such as self-driving vehicles, it is a different story. The components of risk could be the probability of the vehicle laser-light sensors "seeing" a pedestrian, the likelihood of a pedestrian acting predictably, and the chances of a driver taking control at the exact moment when a pedestrian is unseen or unpredictable. The physics of pulsed laser-light sensors is well understood, but how they perform in snow and gloom is not. Research on how pedestrians interact with autonomous vehicles barely exists. And studies of drivers predict that they cannot stay vigilant enough to handle infrequent emergencies.

When scientific understanding is incomplete, risk analysis shifts from reliance on established facts to expert judgment. Studies of those judgments find that they are often quite good–but only when experts get good feedback. For example, meteorologists routinely compare their probability-of-precipitation forecasts with the rain gauge at their station. Given that clear, prompt feedback, when forecasters say that there is a 70 percent chance of rain, it rains about 70 percent of the time. With new technologies such as the self-driving car or gene editing, however, feedback will be a long time coming. Until it does, we will be unsure–and the experts themselves will not know–how accurate their risk estimates really are.

The Science of Climate Science

Expert judgment, which is dependent on good feedback, comes into play when one is predicting the costs and benefits of attempts to slow climate change or to adapt to it. Climate analyses combine the judgments of experts from many research areas, including obvious ones, such as atmospheric chemistry and oceanography, and less obvious ones, such as botany, archaeology and glaciology. In complex climate analyses, these expert judgments reflect great knowledge driven by evidence-based feedback. But some aspects still remain uncertain.

My first encounter with these analyses was in 1979, as part of a project planning the next 20 years of climate research. Sponsored by the Department of Energy, the project had five working groups. One dealt with the oceans and polar regions, a second with the managed biosphere, a third with the less managed biosphere, and a fourth with economics and geopolitics. The fifth group, which I joined, dealt with social and institutional responses to the threat.

Even then, 40 years ago, the evidence was strong enough to reveal the enormous gamble being taken with our planet. Our overall report, summarizing all five groups, concluded that "the probable outcome is beyond human experience."

Thinking of the Unthinkable

How, then, can researchers in this area fulfill their duty to inform people about accurate ways to think about events and choices that are beyond their experience? Scientists can, in fact, accomplish this if they follow two basic lessons from studies of decision-making.

Lesson 1: The facts of climate science will not speak for themselves. The science needs to be translated into terms that are relevant to people's decisions about their lives, their communities and their society. While most scientists are experienced communicators in a classroom, out in the world they may not get feedback on how clear or relevant their messages are.

Addressing this feedback problem is straightforward: test messages before sending them. One can learn a lot simply by asking people to read and paraphrase a message. When communication researchers have asked for such rephrasing about weather forecasts, for example, they have found that some are confused by the statement that there is a "70 percent chance of rain." The problem is with the words, not the number. Does the forecast mean it will rain 70 percent of the time? Over 70 percent of the area? Or there is a 70 percent chance of at least 0.01 inch of rain at the weather station? The last interpretation is the correct answer.

Many studies have found that numbers, such as 70 percent, generally communicate much better than "verbal quantifiers," such as "likely," "some" or "often." One classic case from the 1950s involves a U.S. National Intelligence Estimate that said that "an attack on Yugoslavia in 1951 should be considered a serious possibility." When asked what probability they had in mind, the analysts who signed the document gave a wide range of numbers, from 20 to 80 percent. (The Soviets did not invade.)

Sometimes people want to know more than the probability of rain or war when they make decisions. They want to understand the processes that lead to those probabilities: how things work. Studies have found that some critical aspects of climate change research are not intuitive for many people, such as how scientists can bicker yet still agree about the threat of climate change or how carbon dioxide is different from other pollutants. (It stays in the atmosphere longer.) People may reject the research results unless scientists tell them more about how they were derived.

Lesson 2: People who agree on the facts can still disagree on what to do about them. A solution that seems sound to some can seem too costly or unfair to others.

For example, people who like plans for carbon capture and sequestration, because it keeps carbon dioxide out of the air, might oppose using it on coal-fired power plants. They fear an indirect consequence: cleaner coal may make mountaintop-removal mining more acceptable. Those who know what cap-and-trade schemes are meant to do–create incentives for reducing emissions–might still believe that they will benefit banks more than the environment.

These examples show why two-way communication is so important in these situations. We need to learn what is on others' minds and make them feel like partners in decision-making. Sometimes that communication will reveal misunderstandings that research can reduce. Or it may reveal solutions that make more people happy. One example is British Columbia's revenue-neutral carbon tax, whose revenues make other taxes lower; it has also produced broad enough political support to weather several changes of government since 2008. Sometimes, of course, better two-way communication will reveal fundamental disagreements, and in those cases action is a matter for the courts, streets and ballot boxes.

More than Science

These lessons about how facts are communicated and interpreted are important because climate-related decisions are not always based on what research says or shows. For some individuals, scientific evidence or economic impacts are less important than what certain decisions reveal about their beliefs. These people ask how their choice will affect the way others think about them, as well as how they think about themselves.

For instance, there are people who forgo energy conservation measures but not because they are against conservation. They just do not want to be perceived as eco-freaks. Others who conserve do it more as a symbolic gesture and not based on a belief that it makes a real difference. Using surveys, researchers at Yale Climate Connections have identified what they call Six Americas in terms of attitudes, ranging from alarmed to dismissive. People at those two extremes are the ones who are most likely to adopt measures to conserve energy. The alarmed group's motives are what you might expect. Those in the dismissive group, though, may see no threat from climate change but also have noted they can save money by reducing their energy consumption.

Knowing the science does not necessarily mean agreeing with the science. The Yale study is one of several that found greater polarization among different political groups as people in the groups gained knowledge of some science-related issues. In ongoing research, Caitlin Drummond, currently a postdoctoral fellow at the University of Michigan's Erb Institute, and I have uncovered a few hints that might account for this phenomenon. One possible explanation is that more knowledgeable people are more likely to know the position of their affiliated political group on an issue and align themselves with it. A second possibility is that they feel more confident about arguing the issues. A third, related explanation is that they are more likely to see, and seize, the chance to express themselves than those who do not know as much.

When Decisions Matter Most

Although decision science researchers still have much to learn, their overall message about ways to deal with uncertain, high-stakes situations is optimistic. When scientists communicate poorly, it often indicates that they have fallen prey to a natural human tendency to exaggerate how well others understand them. When laypeople make mistakes, it often reflects their reliance on mental models that have served them well in other situations but that are not accurate in current circumstances. When people disagree about what decisions to make, it is often because they have different goals rather than different facts.

In each case, the research points to ways to help people better understand one another and themselves. Communication studies can help scientists create clearer messages. And decision science can help the public to refine their mental models to interpret new phenomena. By reducing miscommunication and focusing on legitimate disagreements, decision researchers can help society have fewer conflicts and make dealing with the ones that remain easier for us all.

About the Author

Psychologist Baruch Fischhoff is Howard Heinz University Professor in the department of engineering and public policy and the Institute for Politics and Strategy at Carnegie Mellon University. He is a member of the National Academy of Sciences and National Academy of Medicine and past president of the Society for Risk Analysis.

Section 5: Education

Fractions: Where It All Goes Wrong

By Robert S. Siegler

Many children never master fractions. When asked whether 12/13 + 7/8 was closest to 1, 2, 19, or 21, only 24% of a nationally representative sample of more than 20,000 US 8th graders answered correctly. This test was given almost 40 years ago, which gave Hugo Lortie-Forgues and me hope that the work of innumerable teachers, mathematics coaches, researchers, and government commissions had made a positive difference. Our hopes were dashed by the data, though; we found that in all of those years, accuracy on the same problem improved only from 24% to 27% correct.

Such difficulties are not limited to fraction estimation problems nor do they end in 8th grade. On standard fraction addition, subtraction, multiplication, and division problems with equal denominators (e.g., 3/5+4/5) and unequal denominators (e.g., 3/5+2/3), 6th and 8th graders tend to answer correctly only about 50% of items. Studies of community college students have revealed similarly poor fraction arithmetic performance. Children in the US do much worse on such problems than their peers in European countries, such as Belgium and Germany, and in Asian countries such as China and Korea.

This weak knowledge is especially unfortunate because fractions are foundational to many more advanced areas of mathematics and science. Fifth graders' fraction knowledge predicts high school students' algebra learning and overall math achievement, even after controlling for whole number knowledge, the students' IQ, and their families' education and income. On the reference sheets for recent high school AP tests in chemistry and physics, fractions were part of more than half of the formulas. In a recent survey of 2300 white collar, blue collar, and service workers, more than two-thirds indicated that they used fractions in their work. Moreover, in a nationally representative sample of 1,000 Algebra 1 teachers in the

US, most rated as "poor" their students' knowledge of fractions and rated fractions as the second greatest impediment to their students mastering algebra (second only to "word problems").

Why are fractions so difficult to understand? A major reason is that learning fractions requires overcoming two types of difficulty: inherent and culturally contingent. Inherent sources of difficulty are those that derive from the nature of fractions, ones that confront all learners in all places. One inherent difficulty is the notation used to express fractions. Understanding the relation *a/b* is more difficult than understanding the simple quantity *a*, regardless of the culture or time period in which a child lives. Another inherent difficulty involves the complex relations between fraction arithmetic and whole number arithmetic. For example, multiplying fractions involves applying the whole number operation independently to the numerator and the denominator (e.g., 3/7 * 2/7 = (3*2)/(7*7) = 6/49), but doing the same leads to wrong answers on fraction addition (e.g., 3/7 + 2/7 ≠ 5/14). A third inherent source of difficulty is complex conceptual relations among different fraction arithmetic operations, at least using standard algorithms. Why do we need equal denominators to add and subtract fractions but not to multiply and divide them? Why do we invert and multiply to solve fraction division problems, and why do we invert the fraction in the denominator rather than the one in the numerator? These inherent sources of difficulty make understanding fraction arithmetic challenging for all students.

Culturally contingent sources of difficulty, in contrast, can mitigate or exacerbate the inherent challenges of learning fractions. Teacher understanding is one culturally-contingent variable: When asked to explain the meaning of fraction division problems, few US teachers can provide any explanation, whereas the large majority of Chinese teachers provide at least one good explanation. Language is another culturally-contingent factor; East Asian languages express fractions such as 3/4 as "out of four, three," which makes it easier to understand their meaning than relatively opaque terms such as "three fourths." A third such variable is textbooks. Despite division being the most difficult operation to understand, US textbooks present far fewer problems with fraction division than fraction multiplication; the opposite is true in Chinese and Korean textbooks. Probably most fundamental are cultural attitudes: Math learning is viewed as crucial throughout East Asia, but US attitudes about its importance are far more variable.

Given the importance of fractions in and out of school, the extensive evidence that many children and adults do not understand them, and the inherent difficulty of the topic, what is to be done? Considering culturally contingent factors points to several potentially useful steps. Inculcating a deeper understanding of fractions among teachers will likely help them to teach more effectively. Explaining the meaning of fractions to students using clear language (for example, explaining that 3/4 means 3 of the 1/4 units), and requesting textbook writers to include more challenging problems are other promising strategies. Addressing inherent sources of difficulty in fraction arithmetic, in particular understanding of fraction magnitudes, can also make a large difference.

Fraction Face-off!, a 12-week program designed by Lynn Fuchs to help children from low-income backgrounds improve their fraction knowledge, seems especially promising. The program teaches children about fraction magnitudes through tasks such as comparing and ordering fraction magnitudes and locating fractions on number lines. After participating in Fraction Face-off!, fourth graders' fraction addition and subtraction accuracy consistently exceeds that of children receiving the standard classroom curriculum. This finding was especially striking because Fraction Face-off! devoted less time to explicit instruction in fraction arithmetic procedures than did the standard curriculum. Similarly encouraging findings have been found for other interventions that emphasize the importance of fraction magnitudes. Such programs may help children learn fraction arithmetic by encouraging them to note that answers such as 1/3+1/2 = 2/5 cannot be right, because the sum is less than one of the numbers being added, and therefore to try procedures that generate more plausible answers. These innovative curricula seem well worth testing on a wider basis.

About the Author

Robert S. Siegler was Teresa Heinz Professor of Cognitive Psychology at Carnegie Mellon University, and as of September, 2018, Schiff Foundations Professor of Psychology at Columbia University, Teachers College.

Learning on the Back of an Envelope

By Amanda Baker

A few weeks ago, a factoid started bouncing around my social media feeds highlighting the lack of perspective people generally have about large numbers–specifically how much bigger a billion is than a million. The various versions explained that while 1 million seconds is about 11.5 days, 1 billion seconds is about 31.5 years. It's a great example of how scale can slip away from us and how simple calculations can provide perspective. I started looking for other examples and wound up learning just how different other students' introductions to back-of-the-envelope calculations were than my own.

My own introduction came from the arithmetic of paying for lunch. Looking at $5 per day for lunch may seem like an affordable amount–just $5. But what happens when you multiply that by the ~260 work days in a given year? That lunch suddenly costs $1,300 out of your annual budget. Two bucks for coffee? $520. Increase daycare costs by $10 per week–that's another $520. When people want you to join a committee or club and suggest finding a sitter, getting a weekly sitter for three hours at the average ~12/hr will run more than $1,800 for the year. For many families, the back of the envelope is where first-order financial decisions get made.

But when I looked online for similar experiences, most of the stories I found came from college campuses instead of kitchen tables. Stanford holds these kinds of calculations in such high regard that they required all incoming physics majors to enroll in "Science on the Back of an Envelope." Carleton College has a suite of resources for using simple calculations and estimation techniques to introduce geoscience students to time and length scales that might be hard for them to imagine. Whether calculating the cost of diapers or figuring out how to lift James and his Giant Peach with seagulls on strings (an actual assignment in the Stanford course), these calculations take an abstract idea and break it down into chunks. They provide the opportunity to make estimates about relevant details and create a sanity check.

Both science and life have a place for precise calculations, but initial guesstimates provide a place to start. Asking kids to estimate how long it would take to walk to Disneyland opens up a huge question and a series of small assumptions and answers. There are pieces to start with–starting location, a specific distance to Disneyland, hours in a day–but then they have to start breaking it into chunks. How tall is the person walking? How long is their stride? Pick some round numbers that seem reasonable. Or maybe start with an average time it takes them to walk a single mile. How many hours could they go without stopping? Will they sleep or eat? At some point they need to stop adding new details and settle on an estimate that seems good enough.

Practice at breaking down silly or low-stakes problems makes breaking down the bigger ones seem less scary. It also seems more feasible to check something that doesn't seem quite right when you have done it before. When a kid learns to knit and wants to sell scarves, they need to know how to break things down and name a price. Fifteen dollars might sound like a nice amount of money, but a scarf that takes 4 hours to make would be $3.75 per hour without the cost of yarn. How many more bags of dirt would it take to fill a backyard garden, if you increase it by a foot on each side? How many hours per week will a person lose if their daily commute increases by 20 minutes? When people who have had practice with smaller problems read something or hear a story on the news later, they can perform their own sanity check. The huge price tag for a statewide education program may not seem so outrageous when considered in dollars per student, or per taxpayer. They can recognize that comparing millions of dollars for certain things to billions of something else is a great way to hide a huge difference in scale, and they can take a few minutes to break down the math themselves. And–as those same social media feeds later pointed out–when they hear a company is worth a trillion dollars, they can use the fact that a trillion seconds would be more than 31,709 years to gain some perspective.

About the Author

Amanda Baker is a science communicator and outreach advocate. She has a geoscience PhD from Cornell University and has managed open-access, academic journals as well as the outreach journal Frontiers for Young Minds. *She is currently writing and editing science content for kids, from curriculum materials to magazines like* Smore. *She has served as a Science Olympiad national event supervisor and taught a first-year writing seminar on sustainable earth systems while at Cornell.*

The Surprising Link between Recreational Math and Undecidability

By Evelyn Lamb

Julia Robinson was born on December 8, 1919. She was a brilliant mathematician, generous research collaborator, and kind human being who faced some trying circumstances with a great deal of grace. I wrote about her life and work for *Science News* in honor of her 100th birthday.

Most of Robinson's research focused on decision problems: given a particular set of conditions, can one tell whether an arbitrary problem has a solution or not? Specifically, she spent much of her life studying Diophantine equations, polynomials with whole number coefficients, like $x^2+2xy-3y^2=0$. Some Diophantine equations have whole number solutions (in the above, there are many: $x=3$ and $y=3$ is one), and some do not.

Hilbert's 10th problem, one of the challenges David Hilbert issued to the mathematics community in 1900, asked whether there was a universal algorithm that could look at any Diophantine equation and decide whether it had whole number solutions or not. In 1970, Russian mathematician Yuri Matiyasevich built on the work of Robinson did with two other American mathematicians, Martin Davis and Hilary Putnam, to show that no such algorithm could exist. In the limited space of my *Science News* article, I couldn't include the serendipitous way the Fibonacci numbers fit into the ultimate solution of the problem, but it's a fun story, and I wanted to share it here.

The Fibonacci sequence–0, 1, 1, 2, 3, 5, 8, 13, 21,...–is familiar to many math enthusiasts. Starting with the numbers 0 and 1, each term is the sum of the previous two terms. Fibonacci numbers and their cousin the golden ratio are a bit of a recreational math cliché. Frankly, I'm a bit tired of the Fibonacci spiral and Fibonacci numbers in pinecones. But seeing them pop up as part of the solution to

Hilbert's 10th problem, I felt like I was bumping into a familiar friend in a difficult setting.

In a charming article about working with Robinson, Matiyasevich writes about the search for equations that would produce sequences of numbers with specific properties as solutions. The solutions needed to have particular recurrence relations: later solutions were combinations of earlier solutions. Robinson, Davis, and Putnam had shown that finding such an equation would be sufficient to settle Hilbert's 10th problem, and Robinson had zeroed in on finding an equation with the correct properties that would produce the powers of 2. But she couldn't quite get there.

Matiyasevich showed that the Fibonacci numbers could work instead for a modified version of Robinson, Davis, and Putnam's argument. Key to his proof was the following fact: If the square of the nth Fibonacci number divides the mth Fibonacci number, then the nth Fibonacci number divides m. (For this property to work, note that 0 is the 0th term of the Fibonacci sequence and 1 is the 1st term.) For example, the square of the 3rd Fibonacci number, 2, is 4 (so n is 3.) The number 4 divides 8, which is the 6th Fibonacci number (so m is 6). The 3rd Fibonacci number, 2, divides 6, so the property holds. The same thing works for the 4th Fibonacci number, whose square, 9, divides the 12th Fibonacci number, 144. Keeping track of m and n is a little bit challenging, but the property is not particularly difficult to verify for specific examples if you can keep them straight in your mind.

Matiyasevich writes, "It is not difficult to prove this remarkable property of Fibonacci numbers after it has been stated, but it seems that this beautiful fact was not discovered until 1969." He was able to prove it because he was familiar with another theorem in the third edition of a number theory book by Nikolai Vorob'ev. He suspects that if the theorem had been in the earlier edition of Vorob'ev's book that Robinson had had access to, she and her colleagues may have solved the problem a decade earlier.

Matiyasevich describes not only an interesting application of the Fibonacci sequence but also a great example of how mathematics happens. Rarely if ever do lone geniuses work things out all by themselves. His ideas developed through reading other people's work. He found his way to the solution because he was asked to review a paper by Robinson: "I saw at once that Julia Robinson had a fresh and wonderful idea," he writes. It gave him ideas for a new approach to the problem, and his familiarity with Vorob'ev's helped him fill in some gaps. He was part of a community.

After Matiyasevich solved Hilbert's 10th problem, he and Robinson maintained a long-distance collaboration through the mail. They bounced ideas off of each other, caught (and sometimes did not catch) each other's mistakes, and pushed each other forward. Their mathematical friendship bridged differences in age, gender, and nationality. On the centennial of Robinson's birth, I raise a glass to mathematical curiosity, generous coauthors, and the hope that the Fibonacci sequence still has a few surprises in store for us.

About the Author

Evelyn Lamb is a freelance math and science writer based in Salt Lake City, Utah.

To Keep Students in STEM Fields, Let's Weed Out the Weed-Out Math Classes

By Pamela Burdman

All routes to STEM (science, technology, engineering and mathematics) degrees run through calculus classes. Each year, hundreds of thousands of college students take introductory calculus. But only a fraction ultimately complete a STEM degree, and research about why students abandon such degrees suggests that traditional calculus courses are one of the reasons. With scientific understanding and innovation increasingly central to solving 21st-century problems, this loss of talent is something society can ill afford.

Math departments alone are unlikely to solve this dilemma. Several of the promising calculus reforms highlighted in our report *Charting a New Course: Investigating Barriers on the Calculus Pathway to STEM*, published with the California Education Learning Lab, were spearheaded by professors outside of math departments. It's time for STEM faculty to prioritize collaboration across disciplines to transform math classes from weed-out mechanisms to fertile terrain for cultivating a diverse generation of STEM researchers and professionals.

This is not uncharted territory.

In 2013, life sciences faculty at the University of California, Los Angeles, developed a two-course sequence that covers classic calculus topics such as the derivative and the integral, but emphasizes their application in a biological context. The professors used modeling of complex systems such as biological and physiological processes as a framework for teaching linear algebra and a starting point for teaching the basics of computer programming to support students' use of systems of differential equations.

Creating this course, Mathematics for Life Scientists, wasn't easy. The life sciences faculty involved, none of whom had a joint appointment with the math department, said they resorted to designing the course themselves after math faculty rebuffed their overture. The math faculty feared creating a "watered-down" course with no textbook (though after the course was developed, one math instructor taught some sections of the class). Besides math, the life sciences faculty said they experienced "significant pushback" from the chemistry and physics departments over concerns that the course wouldn't adequately prepare students for required courses in those disciplines.

But the UCLA course seems to be successful, and a textbook based on it now exists.

According to recently published research led by UCLA education researchers, students in the new classes ended up with "significantly higher grades" in subsequent physics, chemistry and life sciences courses than students in the traditional calculus course, even when controlling for factors such as demographics, prior preparation and math grades. Students' interest in the subject doubled, according to surveys.

The UCLA example highlights longstanding concerns about the relevance of traditional calculus for biology students. Traditionally, college math departments have overseen general education math courses for students in other majors. A little over two decades ago, biology faculty convened by the Mathematics Association of America advocated that, for biology majors, "statistics, modeling, and graphical representation should take priority" over calculus. But change has been slow, until life sciences departments got involved.

Math education researchers consider more relevant and engaging curriculum to be an important strategy for increasing persistence rates, particularly among students traditionally excluded from STEM fields, such as Black, Latinx and Indigenous students, as well as women.

Engineering departments also worry about calculus sequences driving attrition. In Ohio, Wright State University's solution also involved revising math offerings. But rather than changing the content of the calculus course, they focused on preparing students for calculus by emphasizing "engineering motivation for math." In lieu of traditional calculus prerequisites such as precalculus or college algebra, the engineering faculty launched a contextualized math course in 2004. Emphasizing problem-based learning, the course covers topics students need in sophomore engineering classes, including linear equations, quadratic equations, 2-D vectors and complex numbers. A modest redesign of the engineering curriculum allows students to delay taking a traditional calculus sequence until later in their programs.

The approach enhanced opportunities for students with weaker math backgrounds to succeed in engineering and doubled the average graduation rate of engineering students without reducing the average grade of graduates. Students from groups historically underrepresented in STEM (female, Black, and Latinx), experienced the greatest gains, with those in the new problem-based course completing engineering degrees at three times the rate of engineering majors who did not take it. At least 15 other universities are replicating the strategy.

Increasingly leading math and science organizations are recognizing the importance of interdisciplinary collaboration; the MAA has a history of convening faculty from partner disciplines, and a National Academies' 2013 publication called for reassessing math education in a cross-disciplinary context. The National Science Foundation, which funded both the UCLA and Wright State innovations, recognizes the value of cross-disciplinary or "convergence" research, which is driven by a compelling scientific or societal problem. Low persistence in STEM majors and lack of diversity in the STEM fields are themselves pressing societal problems.

Yet, math departments without jointly-appointed professors seem to be less interested in evidence-based contributions from other disciplines to enhance the effectiveness of math instruction–or even aware of successes to date. The shift toward more practical applications of calculus is missing one key academic endorsement: publication in widely-read journals, if the success of the courses is examined academically at all.

The UCLA research appears in a life sciences education journal, and presentations on Wright State's innovation reach an American Society for Engineering Education audience. Faculty responsible for undergraduate math education are not likely to see these journals.

Ten years ago, at the University of California Berkeley, Lior Pachter used his joint appointments in math and biology to create a new course with the help of his fellow mathematicians and biologists. Methods of Mathematics: Calculus, Statistics, and Combinatorics (Math 10A and 10B), expands on the standard calculus curriculum to reflect how data and technology have transformed the field of biology. Today, most Berkeley life sciences majors take the sequence, and the campus is offering 15 sections of 10B beginning this spring. (Pachter now teaches at the California Institute of Technology, and says Berkeley's math department has yet to publish any research on the calculus sequence, despite having institutionalized it.)

Math learning is fundamental to all STEM fields, but the opposite also appears to be true: the STEM fields may be central to making math learning effective for more students. Involving other STEM disciplines in redesigning math classes is a key way to ensure those classes offer engaging and inclusive on-ramps to STEM.

About the Author

Pamela Burdman is executive director of Just Equations, a policy institute focused on the role of math in education equity.

Section 6: Culture

A Random Walk through the English Language

By Jordan Ellenberg

Here's a game that Claude Shannon, the founder of information theory, invented in 1948. He was trying to model the English language as a random process. Go to your bookshelf, pick up a random book, open it and point to a random spot on the page, and mark the first two letters you see. Say they're I and N. Write down those two letters on your page.

Now, take another random book off the shelf and look through it until you find the letters I and N in succession. Whatever the character *following* "IN" is—say, for instance, it's a space—that's the next letter of your book. And now you take down yet another book and look for an N followed by a space, and once you find one, mark down what character comes next. Repeat until you have a paragraph

"IN NO IST LAT WHEY CRATICT FROURE BIRS GROCID

PONDENOME OF DEMONSTURES OF THE REPTAGIN IS

REGOACTIONA OF CRE"

That isn't English, but it kind of *looks* like English.

Shannon was interested in the "entropy" of the English language, a measure, in his new framework, of how much information a string of English text contains. The Shannon game is a Markov chain; that is, it's a random process where the next step you take depends only on the current state of the process. Once you're at LA, the "IN NO IST" doesn't matter; the chance that the next letter is, say, a B is the probability that a randomly chosen instance of "LA" in your library is followed by a B.

And as the name suggests, the method wasn't original to him; it was almost a half-century older, and it came from, of all things, a vicious mathematical/theological beef in late-czarist Russian math.

There's almost nothing I think of as more inherently intellectually sterile than verbal warfare between true religious believers and movement atheists. And yet, this one time at least, it led to a major mathematical advance, whose echoes have been bouncing around ever since. One main player, in Moscow, was Pavel Alekseevich Nekrasov, who had originally trained as an Orthodox theologian before turning to mathematics. His opposite number, in St. Petersburg, was his contemporary Andrei Andreyevich Markov, an atheist and a bitter enemy of the church. He wrote a lot of angry letters to the newspapers on social matters and was widely known as Neistovyj Andrei, "Andrei the Furious."

The details are a bit much to go into here, but the gist is this: Nekrasov thought he had found a mathematical proof of free will, ratifying the beliefs of the church. To Markov, this was mystical nonsense. Worse, it was mystical nonsense wearing mathematical clothes. He invented the Markov chain as an example of random behavior that could be generated purely mechanically, but which displayed the same features Nekrasov thought guaranteed free will.

A simple example of a Markov chain: a spider walking on a triangle with corners labeled 1, 2, 3. At each tick of the clock, the spider moves from its present perch to one of the other two corners it's connected to, chosen at random. So, the spider's path would be a string of numbers

1, 2, 1, 3, 2, 1, 2, 3, 2, 3, 2, 1 ...

Markov started with abstract examples like this, but later (perhaps inspiring Shannon?) applied this idea to strings of text, among them Alexander Pushkin's poem *Eugene Onegin*. Markov thought of the poem, for the sake of math, as a string of consonants and vowels, which he laboriously cataloged by hand. Letters after consonants are 66.3 percent vowels and 33.7 percent consonants, while letters following vowels are only 12.8 percent vowels and 87.2 percent consonants.

So, you can produce "fake Pushkin" just as Shannon produced fake English; if the current letter is a vowel, the next letter is a vowel with probability 12.8 percent, and if the current letter is a consonant, the next one is a vowel with probability 66.3 percent. The results are not going to be very poetic; but, Markov discovered, they can be distinguished from the Markovized output of other Russian writers. Something of their style is captured by the chain.

Nowadays, the Markov chain is a fundamental tool for exploring spaces of conceptual entities much more general than poems. It's how election reformers identify which legislative maps are brutally gerrymandered, and it's how Google figures out which websites are most important (the key is a Markov chain where at each step you're at a certain website, and the next step is to follow a random link from that site). What a neural net like GPT-3 learns—what allows it to produce uncanny imitation of human-written text—is a gigantic Markov chain that counsels it how to pick the next word after a sequence of 500, instead of the next letter after a sequence of two. All you need is a rule that tells you what probabilities govern the next step in the chain, given what the last step was.

You can train your Markov chain on your home library, or on *Eugene Onegin*, or on the huge textual corpus to which GPT-3 has access; you can train it on anything, and the chain will imitate that thing! You can train it on baby names from 1971, and get:

> Kendi, Jeane, Abby, Fleureemaira, Jean, Starlo, Caming, Bettilia ...

Or on baby names from 2017:

> Anaki, Emalee, Chan, Jalee, Elif, Branshi, Naaviel, Corby, Luxton, Naftalene, Rayerson, Alahna ...

Or from 1917:

> Vensie, Adelle, Allwood, Walter, Wandeliottlie, Kathryn, Fran, Earnet, Carlus, Hazellia, Oberta ...

The Markov chain, simple as it is, somehow captures something of the *style* of naming practices of different eras. One almost experiences it as creative. Some of these names aren't bad! You can imagine a kid in elementary school named "Jalee," or, for a retro feel, "Vensie."

Maybe not "Naftalene," though. Even Markov nods.

About the Author

Jordan Ellenberg is John D. MacArthur Professor of Mathematics at the University of Wisconsin–Madison. His writing has appeared in Slate, *the* Wall Street Journal, *the* New York Times, *the* Washington Post, *the* Boston Globe *and the* Believer. *He is author of the book* Shape: The Hidden Geometry of Information, Biology, Strategy, Democracy, and Everything Else.

Farewell to the Fractional Foot

By Evelyn Lamb

What an exciting year 2019 was for units of measure! (Granted, it doesn't take a lot to be an exciting year for units of measure.) In May, the kilogram got a makeover. The old standard, a piece of platinum-iridium alloy housed at the International Bureau of Weights and Measures in France since 1889, was replaced with a definition based on the fundamental constants of the universe. And in October, in a slightly more provincial matter, the U.S. took an important step to phase out the U.S. survey foot.

The U.S. is nearly alone in the world in resisting the metric system, but underneath the weird mishmash of binary and dozenal and other relationships the customary system uses, our measurements are secretly metric. In 1893, the foot got an official definition based on the meter: 1 foot was defined to be precisely 1200/3937 meters.

I doubt that definition of the foot as a fractional portion of a meter–in particular, a fractional portion whose decimal representation does not terminate–was a dig at the metric system. But in my opinion it is at least ironic that the relationship between feet and meters was fractional instead of decimal.

In 1959, that irony was ironed out...almost. The foot was redefined to be precisely 0.3048 meters. The decimal representation of 1200/3937 begins 0.304800609, so the two definitions differ by only about a 10th of an inch per mile. That new foot was adopted for every purpose except land surveying. The new foot, defined as 0.3048 meters, was called the international foot, and the old definition, 1200/3937 meters, sadly was not called the fractional foot but the U.S. survey foot.

The U.S. survey foot was always intended to be a temporary measure. It was supposed to be replaced with the international foot in 1986. A few states did this, but most states still use the U.S. survey foot, leading to discrepancies between states that add up on the scales that surveying uses. A tenth of an inch per mile is not all that much, but in 1000 miles, that's several feet. If you've ever had trouble with GPS directions because of minor inaccuracies or imprecisions, you can see that a few feet can make a significant difference.

The National Institute of Standards and Technology (NIST) believes that replacing the U.S. survey foot with the international foot will streamline surveying and engineering software and make map projections more accurate.

The U.S. survey foot official changed on December 31, 2022, in tandem with a broader update of the National Spatial Reference System. That update adjusts official latitude, longitude, and altitude measurements at much larger scales than the foot discrepancy, so NIST considers the change to the foot relatively minor compared to these larger changes.

About the Author

Evelyn Lamb is a freelance math and science writer based in Salt Lake City, Utah.

For Math Fans: A Hitchhiker's Guide to the Number 42

By Jean-Paul Delahaye

Everyone loves unsolved mysteries. Examples include Amelia Earhart's disappearance over the Pacific in 1937 and the daring escape of inmates Frank Morris and John and Clarence Anglin from Alcatraz Island in California in 1962. Moreover our interest holds even if the mystery is based on a joke. Take author Douglas Adams's popular 1979 science-fiction novel *The Hitchhiker's Guide to the Galaxy*, the first in a series of five. Toward the end of the book, the supercomputer Deep Thought reveals that the answer to the "Great Question" of "Life, the Universe and Everything" is "forty-two."

Deep Thought takes 7.5 million years to calculate the answer to the ultimate question. The characters tasked with getting that answer are disappointed because it is not very useful. Yet, as the computer points out, the question itself was vaguely formulated. To find the correct statement of the query whose answer is 42, the computer will have to build a new version of itself. That, too, will take time. The new version of the computer is Earth. To find out what happens next, you'll have to read Adams's books.

The author's choice of the number 42 has become a fixture of geek culture. It's at the origin of a multitude of jokes and winks exchanged between initiates. If, for example, you ask your search engine variations of the question "What is the answer to everything?" it will most likely answer "42." Try it in French or German. You'll often get the same answer whether you use Google, Qwant, Wolfram Alpha (which specializes in calculating mathematical problems) or the chat bot web app Cleverbot.

Since the first such school was created in France in 2013 there has been a proliferation of private computer-training institutions in the "42 Network," whose name is a clear allusion to Adams's novels. Today the founding company counts more than 15 campuses in its global network. The number 42 also appears in different forms in the film *Spider-Man: Into the Spider-Verse*. Many other references and allusions to it can be found, for example, in the Wikipedia entry for "42 (number)."

The number 42 also turns up in a whole string of curious coincidences whose significance is probably not worth the effort to figure out. For example:

In ancient Egyptian mythology, during the judgment of souls, the dead had to declare before 42 judges that they had not committed any of 42 sins.

The marathon distance of 42.195 kilometers corresponds to the legend of how far the ancient Greek messenger Pheidippides traveled between Marathon and Athens to announce victory over the Persians in 490 B.C. (The fact that the kilometer had not yet been defined at that time only makes the connection all the more astonishing.)

Ancient Tibet had 42 rulers. Nyatri Tsenpo, who reigned around 127 B.C., was the first. And Langdarma, who ruled from 836 to 842 A.D. (i.e., the 42nd year of the ninth century), was the last.

The Gutenberg Bible, the first book printed in Europe, has 42 lines of text per column and is also called the "Forty-Two-Line Bible."

According to a March 6, 2020 *Economist* blog post marking the 42nd anniversary of the radio program *The Hitchhiker's Guide to the Galaxy*, which preceded the novel, "the 42nd anniversary of anything is rarely observed."

A Purely Arbitrary Choice

An obvious question, which indeed has been asked, is whether the use of 42 in Adams's books had any particular meaning for the author. His answer, posted in the online discussion group alt.fan.douglas-adams, was succinct: "It was a joke. It had to be a number, an ordinary, smallish number, and I chose that one. Binary representations, base thirteen, Tibetan monks are all complete nonsense. I sat at my desk, stared into the garden and thought '42 will do.' I typed it out. End of story."

In the binary system, or base 2, 42 is written as 101010, which is pretty simple and, incidentally, prompted a few fans to hold parties on October 10, 2010 (10/10/10). The reference to base 13 in Adams's answer requires a more indirect explanation. In one instance, the series suggests that 42 is the answer to the question "What do you get if you multiply six by nine?" That idea seems absurd because $6 \times 9 = 54$. But in base 13, the number expressed as "42" is equal to $(4 \times 13) + 2 = 54$.

Apart from allusions to 42 deliberately introduced by computer scientists for fun and the inevitable encounters with it that crop up when you poke around a bit in history or the world, you might still wonder whether there is anything special about the number from a strictly mathematical point of view.

Mathematically Unique?

The number 42 has a range of interesting mathematical properties. Here are some of them:

The number is the sum of the first three odd powers of two—that is, $2^1 + 2^3 + 2^5 = 42$. It is an element in the sequence $a(n)$, which is the sum of n odd powers of 2 for $n > 0$. The sequence corresponds to entry A020988 in The On-Line Encyclopedia of Integer Sequences (OEIS), created by mathematician Neil Sloane. In base 2, the nth element may be specified by repeating 10 n times (1010 ... 10). The formula for this sequence is $a(n) = (2/3)(4^n - 1)$. As n increases, the density of numbers tends toward zero, which means that the numbers belonging to this list, including 42, are exceptionally rare.

The number 42 is the sum of the first two nonzero integer powers of six—that is, $6^1 + 6^2 = 42$. The sequence $b(n)$, which is the sum of the powers of six, corresponds to entry A105281 in OEIS. It is defined by the formulas $b(0) = 0$, $b(n) = 6b(n - 1) + 6$. The density of these numbers also tends toward zero at infinity.

Forty-two is a Catalan number. These numbers are extremely rare, much more so than prime numbers: only 14 of the former are lower than one billion. Catalan numbers were first mentioned, under another name, by Swiss mathematician Leonhard Euler, who wanted to know how many different ways an n-sided convex polygon could be cut into triangles by connecting vertices with line segments. The beginning of the sequence (A000108 in OEIS) is 1, 1, 2, 5, 14, 42, 132.... The nth element of the sequence is given by the formula $c(n) = (2n)! / (n!(n + 1)!)$. And like the two preceding sequences, the density of numbers is null at infinity.

Catalan numbers are named after Franco-Belgian mathematician Eugène Charles Catalan (1814–1894), who discovered that $c(n)$ is the number of ways to arrange n pairs of parentheses according to the usual rules for writing them: a parenthesis is never closed before it has been opened, and one can only close it when all the parentheses that were subsequently opened are themselves closed.

For example, $c(3) = 5$ because the possible arrangements of three pairs of parentheses are:

((( ))); ( ) ( ) ( ); (( )) ( ); (( ) ( )); ( ) (( ))

Forty-two is also a "practical" number, which means that any integer between 1 and 42 is the sum of a subset of its distinct divisors. The first practical numbers are 1, 2, 4, 6, 8, 12, 16, 18, 20, 24, 28, 30, 32, 36, 40, 42, 48, 54, 56, 60, 64, 66 and 72 (sequence A005153 in OEIS). No simple known formula provides the nth element of this sequence.

All this is amusing, but it would be wrong to say that 42 is really anything special mathematically. The numbers 41 and 43, for example, are also elements of many sequences. You can explore the properties of various numbers on Wikipedia.

What makes a number particularly interesting or uninteresting is a question that mathematician and psychologist Nicolas Gauvrit, computational natural scientist Hector Zenil and I have studied, starting with an analysis of the sequences in the OEIS. Aside from a theoretical connection to Kolmogorov complexity (which defines the complexity of a number by the length of its minimal description), we have shown that the numbers contained in Sloane's encyclopedia point to a shared mathematical culture and, consequently, that OEIS is based as much on human preferences as pure mathematical objectivity.

Problem of the Sum of Three Cubes

Computer scientists and mathematicians recognize the appeal of the number 42 but have always thought that it was a simple game that could be played just as well with another number. Still, a recent news item caught their attention. When it was applied to the "sum of three cubes" problem, 42 was more troublesome than all the other numbers below 100.

The problem is stated as follows: What integers n can be written as the sum of three whole-number cubes ($n = a^3 + b^3 + c^3$)? And for such integers, how do you find a, b and c? As a practical matter, the difficulty in making this calculation is that for a given n, the space of the triplets to be considered involves negative integers. This triplet space is therefore infinite, unlike the computation for the sum of squares. For that particular problem, any solution has an absolute value lower than the square root of a given n. Moreover for the sum of squares, we know perfectly well what is possible and impossible.

For the sum of cubes, some solutions may be surprisingly large, such as the one for 156, which was discovered in 2007:

$$156 = 26{,}577{,}110{,}807{,}569^3 + (-18{,}161{,}093{,}358{,}005)^3 + (-23{,}381{,}515{,}025{,}762)^3$$

Note that for some integer values of n, the equation $n = a^3 + b^3 + c^3$ has no solution. Such is the case for all integers n that are expressible as $9m + 4$ or $9m + 5$ for any integer m (e.g., 4, 5, 13, 14, 22, 23). Demonstrating this assertion is straightforward: we use the "modulo 9" (mod 9) calculation, which is equivalent to assuming that $9 = 0$ and then manipulating only numbers between 0 and 8 or between -4 and 4. When we do so, we see that:

$0^3 = 0 \pmod 9$; $1^3 = 1 \pmod 9$; $2^3 = 8 = -1 \pmod 9$; $3^3 = 27 = 0 \pmod 9$; $4^3 = 64 = 1 \pmod 9$; $5^3 = (-4)^3 = -64 = -1 \pmod 9$; $6^3 = (-3)^3 = 0 \pmod 9$; $7^3 = (-2)^3 = 1 \pmod 9$; $8^3 = (-1)^3 = -1 \pmod 9$

In other words, the cube of an integer modulo 9 is -1 (= 8), 0 or 1. Adding any three numbers among these numbers gives:

$0 = 0 + 0 + 0 = 0 + 1 + (-1)$; $1 = 1 + 0 + 0 = 1 + 1 + (-1)$; $2 = 1 + 1 + 0$; $3 = 1 + 1 + 1$; $6 = -3 = (-1) + (-1) + (-1)$; $7 = -2 = (-1) + (-1) + 0$; $8 = -1 = (-1) + 0 + 0 = 1 + (-1) + (-1)$

You cannot get a sum of 4 or 5 (= -4). This restriction means that sums of three cubes are never numbers of the form $9m + 4$ or $9m + 5$. We thus say that $n = 9m + 4$ and $n = 9m + 5$ are prohibited values.

Searching for Solutions

To illustrate how difficult it is to find solutions to the equation $n = a^3 + b^3 + c^3$, let's see what happens for $n = 1$ and $n = 2$.

For $n = 1$, there is the obvious solution:

$$1^3 + 1^3 + (-1)^3 = 1$$

Are there others? Yes, there is:

$$9^3 + (-6)^3 + (-8)^3 = 729 + (-216) + (-512) = 1$$

That calculation is not the only other solution. In 1936 German mathematician Kurt Mahler proposed an infinite number of them. For any integer p:

$$(9p^4)^3 + (3p - 9p^4)^3 + (1 - 9p^3)^3 = 1$$

This proposition may be proved using the remarkable identity:

$$(A + B)^3 = A^3 + 3A^2B + 3AB^2 + B^3$$

An infinite set of solutions is also known for $n = 2$. It was discovered in 1908 by mathematician A. S. Werebrusov. For any integer p:

$$(6p^3 + 1)^3 + (1 - 6p^3)^3 + (-6p^2)^3 = 2$$

By multiplying each term of these equations by the cube of an integer ($r3$), we deduce that there are also infinitely many solutions for both the cube and double the cube of any integer.

Consider the example of 16, which is double the cube of 2. For $p = 1$, we get:

$$14^3 + (-10)^3 + (-12)^3 = 16$$

Note that for $n = 3$, as of August 2019, only two solutions were known:

$$1^3 + 1^3 + 1^3 = 3; 4^3 + 4^3 + (-5)^3 = 3$$

A question that naturally follows is: Is there at least one solution for every nonprohibited value?

Computers at Work

To answer that question, mathematicians started by taking the nonprohibited values 1, 2, 3, 6, 7, 8, 9, 10, 11, 12, 15, 16 ... (A060464 in OEIS) and examining them one by one. If solutions can be found for all those examined values, it will be reasonable to conjecture that for any integer n that is not of the form $n = 9m + 4$ or $n = 9m + 5$, there are solutions to the equation $n = a^3 + b^3 + c^3$.

The research carried out thus far, which depends on the power of the computers or computer networks used, has produced an ever expanding body of results. This work leads us back to the famous and intriguing number 42.

In 2009, employing a method proposed by Noam Elkies of Harvard University in 2000, German mathematicians Andreas-Stephan Elsenhans and Jörg Jahnel explored all the triplets a, b, c of integers with an absolute value less than 10^{14} to find solutions for n between 1 and 1,000. The paper reporting their findings concluded that the question of the existence of a solution for numbers below 1,000 remained open only for 33, 42, 74, 114, 165, 390, 579, 627, 633, 732, 795, 906, 921 and 975.

For integers less than 100, just three enigmas remained: 33, 42 and 74.

In a 2016 preprint paper, Sander Huisman, now at the University of Twente in the Netherlands, pressed on and found a solution for 74:

$$(-284{,}650{,}292{,}555{,}885)^3 + (66{,}229{,}832{,}190{,}556)^3 + (283{,}450{,}105{,}697{,}727)^3$$

In 2019 Andrew Booker of the University of Bristol in England settled the case of 33:

$$8{,}866{,}128{,}975{,}287{,}528)^3 + (-8{,}778{,}405{,}442{,}862{,}239)^3 + (-2{,}736{,}111{,}468{,}807{,}040)^3$$

From that point, Douglas Adams's number was the last positive integer lower than 100 whose representation as a sum of three integer cubes was unknown. If there was no solution, that conclusion would provide a genuinely compelling rationale for the mathematical significance of 42: it would be the first number for which a solution appeared possible but none had been found. Computers tried but had been unable to crack the problem.

The answer came in a 2020 preprint, the result of a huge computational effort coordinated by Booker and Andrew Sutherland of the Massachusetts Institute of Technology. Computers participating in the Charity Engine network of personal computers, calculating for the equivalent of more than one million hours, showed:

$$42 = (-80{,}538{,}738{,}812{,}075{,}974)^3 + 80{,}435{,}758{,}145{,}817{,}515^3 + 12{,}602{,}123{,}297{,}335{,}631^3$$

The cases of 165, 795 and 906 were also solved recently. For integers below 1,000, only 114, 390, 579, 627, 633, 732, 921 and 975 remain to be solved.

The conjecture that solutions exist for all integers n that are not of the form $9m + 4$ or $9m + 5$ would appear to be confirmed. In 1992 Roger Heath-Brown of the University of Oxford proposed a stronger conjecture stating that there are infinitely many ways to express all possible n's as the sum of three cubes. The work is far from over.

The difficulty appears so daunting that the question "Is n a sum of three cubes?" may be undecidable. In other words, no algorithm, however clever, may be able to process all possible cases. In 1936, for example, Alan Turing showed that no algorithm can solve the halting problem for every possible computer program. But here we are in a readily describable, purely mathematical domain. If we could prove such undecidability, that would be a novelty.

The number 42 was difficult, but it is not the final step!

About the Author

Jean-Paul Delahaye is a professor emeritus of computer science at the University of Lille in France and a researcher at the Research Center in Computer Science, Signal and Automatics of Lille (CRIStAL). He recently published Les Mathématiciens Se Plient au Jeu (Belin, 2017), a French collection of articles from Pour la Science.

Forget Meters and Feet

By Evelyn Lamb

Most of us learn some amount of Euclidean geometry in school. We learn that the interior angles of triangles add up to 180°. We learn how to prove that lines are parallel, or shapes are congruent or similar. Even though we live on a planet that is not flat, our everyday intuition is on a scale that makes us feel like Euclidean geometry is the natural way to think about shapes, lengths, and angles.

I think it's a real shame that more students are not exposed to non-Euclidean geometry early in their educations, but that's a column for another time. Suffice it to say that if one is lucky enough to encounter geometry beyond that which takes place on a perfectly flat plane, one will learn that there is much more to geometry than two-column proofs and the Pythagorean theorem.

The intuition we develop in Euclidean geometry does not prepare us well for non-Euclidean geometry. One of the delights I found when I first started studying hyperbolic geometry (one of the flavors of non-Euclidean geometry) was that many things that seem so obvious as not to require any kind of justification are flat-out wrong when we leave the flat Euclidean plane.

For example, in non-Euclidean geometry, there is no such thing as a pair of triangles that are similar but not congruent. The geometry of a sphere is another flavor of non-Euclidean geometry, so we can think about it on a globe. On the Earth, there is a triangle that connects the North Pole with Quito (the capital of Ecuador) and Libreville (the capital of Gabon). This triangle is close to being a triangle with three right angles, or 270° of internal angle. (The Atlantic Ocean is growing, so in a few billion years, it will be even closer.) On the other hand, the "golden triangle" in Texas (a name only Denton seems to use for the region spanned by Dallas, Fort Worth, and Denton), has a total internal angle of 180.002°, just a smidge more than a flat triangle. (Thanks to David Radcliffe for designing the tool I used to figure that out.)

If you play with more triangles on a globe (or a grapefruit), you will find that the total area of the triangle increases proportionally to the total angle of the triangle. You can't find a golden triangle-shaped triangle on the earth with cities that are thousands of miles apart. Spherical geometry won't let you do that. In hyperbolic geometry the relationship goes the opposite way: the larger the triangle, the less total angle it has.

In Euclidean geometry, there is no natural measuring stick. Over the course of history, humans have chosen many different ways to measure flat distances. All of them are somewhat arbitrary. And in fact, I think it's in part our experience with flat geometry that makes us think it's normal for units of measurement to be arbitrary.

But it doesn't have to be that way. In spherical and hyperbolic geometry, the formulas that relate the size of a shape with its internal angle always have these pesky coefficients that pop out. The formula for the area of a spherical triangle is $K^2(a+b+c-180)$, where *a*, *b*, and *c* are the measurements (in degrees) of the three angles of the triangle, and the area of a hyperbolic triangle is K2(180−a−b−c). The K^2 factor comes from the intrinsic curvature of the space, or how the space is bending at every point. On a sphere, that curvature is related to the radius of the sphere. A smaller sphere has higher curvature, and a bigger sphere has lower. (A baseball is more curved than a basketball.)

To most mathematicians, that K^2 is a bit distracting. Two spheres of different sizes have more in common than not, and why should you have to keep track of these coefficients? It's a recipe for arithmetic errors. Wouldn't it be better if K^2 were just 1 no matter what? Good news: This paradise can be yours! Just set your unit of measurement to be the radius of the sphere you're looking at. Instead of meters, feet, furlongs, or fathoms, the correct base unit of length for people who live on the earth is the radius of the earth itself.

History of metrology buffs may point out that the meter was originally defined in terms of the size of the earth. Specifically, a meter was defined to be a 10 millionth of the distance between the Equator and North Pole along a line of longitude. (The meter has since been redefined in terms of the speed of light in a vacuum.) But come on, to relate it to a quarter of the circumference and not the radius? That's a real misstep, if you ask me. So I think we need to scrap the meter and start over with the Radius (abbreviated R).

The radius of the earth is about 3,963 miles, so we will need to get really comfortable with scientific notation to start expressing everyday measurements in terms of Radii. I'm 2.589×10^{-7} R tall. I live about 0.25 R away from where I grew up. I'm working on a sewing project that requires me to make a lot of 3.983×10^{-9} R bias tape. I'm sure you'll agree that the extra work is minor compared to the satisfaction of knowing you are using the only correct unit of length the planet has seen fit to provide us with.

"But Evelyn," you say. "The earth is not a sphere."

I take a deep breath. "Frankly, there's no way all the governments on the planet could cooperate effectively enough to maintain a massive conspiracy about the shape of the earth for mi–"

"No, no," you laugh. "I mean, it's an oblate spheroid. It's not a perfect sphere."

Fair enough. It's true, the Earth is not completely spherical. It bulges out slightly at the equator (more than a billiard ball, despite what you may have heard), making it an oblate spheroid rather than a sphere. We've got a couple of options here. First, we could not worry about it. The difference between the largest and smallest distances between the center of the earth and a point on the surface is 13 miles out of almost 4,000, about 0.3%. Nothing we're measuring could be so important that we need to be off by 1 part in 300, could it? The other option is to just pick one value and go with it. The International Union of Geodesy and Geophysics has a few options for radius of the earth, including the equatorial radius (3963 miles) and three different average radii, all of which are pretty close to 3959 miles.

No matter what the final decision is on the precise definition, we can rest assured that the earth has provided us with a beautiful absolute unit and we'll never have to argue about how to measure anything ever again.

About the Author

Evelyn Lamb is a freelance math and science writer based in Salt Lake City, Utah.

Math Is More Than Just Numbers: Celebrate Pi Day a Different Way

By Alissa S. Crans

This Pi Day, as you're enjoying your slice of pie, explore a bigger slice of mathematics.

Most March 14 celebrations are obsessed with the number pi. If you think "obsessed" is too strong, take a look around: you'll see things on sale for $3.14, Pi-K runs and, perhaps as in the past, a segment on daytime TV.

As a mathematician, I'm tickled that honoring something mathematical has become a widespread phenomenon. But, at the same time, I'm disappointed that this numerical celebrity seems to be somewhat of an accident. There is *so much more* to math than one solitary number, regardless of however many digits you can recite from memory.

There's no reason why pi, of all the infinitely many irrational numbers, should receive special treatment. You may remember from high school that numbers are either rational or irrational and what distinguishes them from one another. Rational numbers can be expressed in one or more of the following ways:

- fractions, such as 7/8 or 3/1
- decimals that repeat, even if forever, such as 1.99999... or 2.488488488...
- decimals that terminate, such as .123456789

Irrationals, on the other hand, can't be written in any of these ways. Pi, for example, never terminates, so we throw up our hands and use the common approximation of 3.14. What most of us don't learn is that there are many more irrational numbers than rational numbers, so many more that if you asked a genie to choose a number truly at random, the likelihood it would pick an irrational number is 100 percent!

This means that any other day of the year would do just as well as March 14 if our only goal was to honor a lone number. For example, we might have chosen to latch onto the first number known to be irrational and celebrate the square root of 2, roughly 1.4, on January 4. Or, we could have decided to commemorate 2.7-ish, the number e, which some of you might know from your calculus courses, on February 7. Conceivably, pi won out simply because of its familiarity, because March 14 enables us to use three of its digits, and because phonetically it shares the name of a tasty dessert. Like a lot of people, I'm happy to have an excuse to indulge in a slice of my favorite pie, apple crumb.

But why just focus on numbers when there's so much more beauty, wonder and excitement to be found in math? The International Mathematical Union shares this sentiment. Thanks to them, March 14 has become the annual International Day of Mathematics, enabling us to celebrate all of mathematics as opposed to just a single number. To whet your appetite, here's a small sample of things you might be surprised that math has something to say about: cutting cake, juggling, performing magic tricks, tiling your bathroom floor or creating knitting patterns.

Here's what I mean: the next time you're celebrating a birthday, you might want to make sure everyone gets the same amount of cake and frosting. Math can tell you what shapes your cake must be to make this possible.

Intrigued? Fortunately, there are numerous ways to start exploring the diversity of math. Videos from Standupmaths, Numberphile and 3Blue1Brown offer entertaining snapshots of fun math topics you might not have seen before. Or, if you'd like to involve your kids, there's an award called the Mathical for books "that inspire children of all ages to see math in the world around them." You might choose one of the prize winners to read together.

Or maybe you're the type that likes to do and not just watch or read. The National Museum of Mathematics offers an incredible array of events and activities for all ages. There are also many places to find intriguing problems and questions, with no special background required, such as the Riddler on FiveThirtyEight and Alex Bellos's Monday Puzzle in the *Guardian*.

Or, look up Martin Gardner, the undisputed master of presenting entertaining and thought-provoking puzzles. Try your hand at one of his many "Mathematical Games" from past issues of *Scientific American*.

But, be careful, or you might just get hooked.

About the Author

Alissa S. Crans is a professor of mathematics at Loyola Marymount University and a fellow of the Association for Women in Mathematics. She has been nationally recognized for her enthusiastic ability to share and communicate mathematics and honored by the Mathematical Association of America with awards for distinguished teaching and expository writing on mathematics.

Mathematicians Are Trying to "Hear" Shapes

By Rachel Crowell

More than 50 years ago Polish-American mathematician Mark Kac popularized a zany but mathematically deep question in his 1966 paper "Can One Hear the Shape of a Drum?" In other words, if you hear someone beat a drum, and you know the frequencies of the sounds it makes, can you work backward to figure out the shape of the drum that created those sounds? Or can more than one drum shape create the exact same set of frequencies?

Kac wasn't the first person to pose this or related questions, but he garnered considerable attention for the subject. In 1968 he won the Mathematical Association of America's Chauvenet Prize, which is focused on mathematical exposition, for his 1966 paper. "It's really well written and very widely accessible," says Julie Rowlett, a mathematician at Chalmers University of Technology in Sweden.

Kac's work pushed these problems, which fall into a mathematical field called isospectral geometry, further into the limelight, inspiring researchers to ask similar questions for different shapes and surfaces. Their work ignited an area of research that is still active and growing today.

Listening to Drums

More than 20 years after Kac's paper, three mathematicians proved that you can't actually hear the shape of a drum. The team was able to produce multiple examples of drums with different geometries that created the same frequencies of sounds.

The researchers' findings began to crystallize in a new way while one of the mathematicians–Carolyn Gordon, now an emeritus professor at Dartmouth College–was on a short visit to Europe. She had traveled to Germany's Mathematical Research Institute of Oberwolfach, nestled in the Black Forest. Despite all the advantages of sojourning to an "idyllic place where you're very far from everything," Gordon says, her time at Oberwolfach "just happened to be the week when things were falling into place" for the team's research on hearing shapes.

She had been working on related problems for years. Gordon's doctoral thesis involved studying how to discern "whether two shapes that are presented in a sort of abstract manner" are the same, she says. Through this other research question, she "slipped into" working on the drum problem.

But the institute wasn't set up for visitors to reach the outside world easily. "There was a phone that you could use at certain hours at night, but you had to wait in line," Gordon says. "It was challenging to connect, but it was an exciting time." One of the other mathematicians on her team, David Webb, who is married to Gordon, shares similar memories. "We were trying to settle the question as quickly and efficiently as we could, because the question had been open for quite some time, and we were eager to get something written up in print," says Webb, now a mathematician at Dartmouth.

A turning point happened when the researchers realized that an example Gordon previously thought wasn't going to work was just what they needed to show two differently shaped drums that sound identical. "We got ideas for other pairs that were much more complicated. We were making these huge paper constructions" to represent drums of different shapes, and then "trying to smash them," she says. After creating those paper "monstrosities," as Webb called them, the mathematicians found out they didn't work. "And then we went back to the original pair and realized it was fine," Gordon says.

Effectively, their work had answered a question that earlier researchers considered intractable. In 1882 Arthur Schuster, a German-born British physicist, wrote, "To find out the different tunes sent out by a vibrating system is a problem which may or may not be solvable in certain special cases, but it would baffle the most skilful [sic] mathematician to solve the inverse problem and to find out the shape of a bell by means of the sounds which it is capable of sending out."

The discovery was a major step but still left many questions unanswered.

The Rule or the Exception

In the past several decades, researchers have solved a host of problems about "hearing" the sounds of shapes.

It turns out you can hear the shape of a triangle, a result first proved in Catherine Durso's 1988 doctoral thesis for the Massachusetts Institute of Technology. You can also hear the shape of parallelograms and acute trapezoids, according to a 2015 paper by Rowlett and Zhiqin Lu, a mathematician at the University of California, Irvine. Both shapes produce unique sounds. And that paper yielded additional interesting findings, Rowlett explains.

"Let's say you're making quadrilateral drums, so four straight edges," she says. "You would be able to hear a square one. It would sound special. And the same thing for triangle drums: an equilateral triangle drum would sound special, not like any of the others." Moreover, for any regular polygonal drum–a shape with equal side lengths and equal interior angles–"you would always be able to hear it amongst the others. And I like to think it would sound particularly nice," Rowlett says.

You can also hear the shape of a truncated cone–that is, a cone that has its pointy tip cut off, researchers reported in the December 2021 issue of *Physical Review E*. Also in 2021 Rowlett and her colleagues showed that you can discern the shape of a trapezoid from sounds if it isn't obtuse.

Yet among all the individual results about hearing shapes, a different team of researchers pointed out a gaping unsettled idea: it remains to be seen whether it is generally true that you will be able to discern the outline of a given type of shape or surface from its sounds.

The question of the relationship between a shape and its associated set of frequencies "is far from being closed, from both theoretical and practical perspectives," researchers wrote in a 2018 paper presented at the IEEE/CVF Conference on Computer Vision and Pattern Recognition. "Specifically, it is not yet certain whether the counterexamples," such as the case of the drum, "are the rule or the exception. So far, everything points towards the latter."

Some of the questions concerning "hearing" shapes have taken researchers to places that are challenging to even picture: higher dimensions.

Visiting Fantastical Dimensions

One of Rowlett's recent preprint papers connects to a problem that was solved way back in 1964 by mathematician John Milnor, now at Stony Brook University. It involves journeying beyond the familiar three dimensions of space to a difficult-to-envision mathematical realm of 16 dimensions.

"We're thinking about [flat] tori," Rowlett says. In one dimension, a torus "is just a circle," she notes. In three dimensions, mathematicians often describe tori as having the shape of a glazed doughnut, though they're usually only referring to the surface of the sugary delight, not its doughy innards.

But Milnor considered what happens when one listens to the shapes of even more mysterious and abstract surfaces: 16-dimensional tori. He found, basically, that one can't hear the shape of tori in 16 dimensions.

It might seem odd to jump to the 16th dimension, but there are surprisingly practical reasons for doing so. "The more dimensions you have, the more ways there are for things to be geometrically different," Rowlett says. Thus, this case was actually "a simple example where it was easy to see" those differences, she notes.

Milnor's paper, which is just one page long, "inspired Kac to a great extent. So that was a fundamental contribution to getting this field growing," Rowlett says. But Milnor's work left open the question of whether one can hear the shape of lower-dimensional flat tori. "What about 15-dimensional–or 14?" Rowlett asks.

Rowlett's recent preprint paper, which she co-authored with two researchers who were then her students, was motivated by her desire to discover "the tipping point" between when you can and can't hear the shape of a flat torus. "Three is the magic number," meaning one can't hear the shape of tori in four or more dimensions, she says.

But reaching that answer required Rowlett's team to take a circuitous path. Surprisingly, her then students Erik Nilsson and Felix Rydell discovered that the question had already been answered. But the problem's solution was buried in work from the 1990s by mathematician Alexander Schiemann.

The connection between Schiemann's work and the question Rowlett was pondering was so muffled by mathematical differences that it had escaped wider recognition. That is largely because the answer to the question "was published entirely using number theory language," she says. Key words such as "isospectral" weren't mentioned. "The paper that proves this mathematically never even mentions the word 'torus,'" she notes.

Therefore, in their not yet published paper, Rowlett, Nilsson and Rydell provide three mathematical perspectives–analytic, geometric and number theoretical–on the problem Schiemann studied, building bridges that connect the technical aspects of understanding his results from the three mathematical viewpoints.

"People who are interested in these types of problems then have access, as well, to the tools from the different fields," Rowlett says. Maybe now, when a different team needs to pull out a related result, they won't have to dig so deep looking for it, she says.

Amplifying Mathematics

In the late 1800s, when Schuster mused about the immense challenge of determining the shape of a bell by the sounds it emits, microphones were a new technology. More than 130 years later a team of researchers used microphones in a way that might have shocked Schuster. They employed them to show that you can, in a sense, hear the shapes of rooms–specifically convex, polyhedral ones.

Using a few microphones arranged in an arbitrary setup, the researchers' computer algorithm "reconstructs the full 3D geometry of the room from a single sound emission," they wrote in a 2013 paper. The scientists noted that their findings could be applied to problems in architectural acoustics, virtual reality, audio forensics, and more.

The landscape of research surrounding hearing different shapes and surfaces has shifted considerably since Schuster's time. With the continued meeting of mathematical minds from different fields and additional advances in technology, who knows what new sounds and shapes mathematicians will explore in coming decades.

About the Author

Rachel Crowell is a Midwest-based writer covering science and mathematics.

GLOSSARY

aggregate Formed into a collection of mass of things.

anthropogenic Human-caused, usually when referring to climate change.

axiom An established, agreed-upon truth or principle.

conjecture In mathematics, an idea or principle that is not yet proved or disproved.

conundrum A difficult problem, especially one with only a speculative answer.

cryptography In computers, the encoding and decoding of information.

disenfranchise To deprive (someone) of a right, especially the right to vote.

economy of scale The increased efficiency and lower cost of goods as the rate of producing them increases.

emphysema A lung condition in which alveoli (air sacs) are damaged, often as a result of smoking.

epistemology The study of how we know what we know.

hypothesis An initial assumption made as a starting point before testing for accuracy.

proof A demonstration that uses logic to determine the truth of a statement.

quandary The state of uncertainty, confusion, or doubt.

quintile In statistics, one-fifth, or 20 percent of any group.

theorem In mathematics, a formula or statement deduced from other formulas or statements.

FURTHER INFORMATION

Boghosian, Bruce, "Is Inequality Inevitable?" *Scientific American*, 1 November 2019, https://www.scientificamerican.com/article/is-inequality-inevitable/

Duchin, Moon, "Geometry versus Gerrymandering," *Scientific American*, 1 November 2018, https://www.scientificamerican.com/article/geometry-versus-gerrymandering/

Horgan, John, "Is Mathematics, like Science, Pluralistic?," *Scientific American*, 14 September 2019, https://blogs.scientificamerican.com/cross-check/is-mathematics-like-science-pluralistic/

Houston-Edwards, Kelsey, "The Mathematics of How Connections Become Global," *Scientific American*, 1 April 2021, https://www.scientificamerican.com/article/the-mathematics-of-how-connections-become-global/

Hullman, Jessica, "How to Get Better at Embracing Unknowns," *Scientific American*, 1 September 2019, https://www.scientificamerican.com/article/how-to-get-better-at-embracing-unknowns/

Lamb, Evelyn, "Computation in Service of Poetry," *Scientific American*, 19 February 2020, https://blogs.scientificamerican.com/roots-of-unity/computation-in-service-of-poetry/

Montgomery, Richard, "The Three-Body Problem," *Scientific American*, 1 August 2019, https://www.scientificamerican.com/article/the-three-body-problem/

CITATIONS

1.1 Hilbert Walked so the Clay Mathematics Institute Could Run by Evelyn Lamb (October 17, 2019); 1.2 Is the Mathematical World Real? by Kelsey Houston-Edwards (September 1, 2019); 1.3 Is the Schrödinger Equation True? by John Horgan (January 7, 2021); 1.4 The Evolving Quest for a Grand Unified Theory of Mathematics by Rachel Crowell (March 21, 2022); 1.5 The Top Unsolved Questions in Mathematics Remain Mostly Mysterious by Rachel Crowell (May 28, 2021); 2.1 A Deep Math Dive into Why Some Infinities Are Bigger Than Others by Martin Goldstern and Jakob Kellner (August 16, 2021); 2.2 Gnarly, Centuries-Old Mathematical Quandaries Get New Solutions by Rachel Crowell (October 14, 2021); 2.3 New Proof Solves 80-Year-Old Irrational Number Problem by Leila Sloman (September 16, 2019); 2.4 The Very Special Triangles by Evelyn Lamb (March 6, 2020); 3.1 Ask Gini: How to Measure Inequality by Evelyn Lamb (November 12, 2012); 3.2 Can an Algorithm Help Solve Political Paralysis? by Julia Hotz (September 18, 2020); 3.3 Simple Mathematical Law Predicts Movement in Cities around the World by Viviane Callier (October 1, 2021); 3.4 Smartphone Data Show Voters in Black Neighborhoods Wait Longer by Daniel Garisto (October 1, 2019); 3.5 The American Economy Is Rigged by Joseph E. Stiglitz (November 1, 2018); 4.1 Algorithm Aids Search for Those Lost at Sea by Scott Hershberger (September 1, 2020); 4.2 Color-Changing Fibers Unravel a Knotty Mystery by Jeremy Hsu (January 3, 2020); 4.3 Math Can Help Build a Global Digital Community by Cindy Lawrence (March 26, 2021); 4.4 The Mathematics of (Hacking) Passwords by Jean-Paul Delahaye (August 5, 2019); 4.5 Why History Should Matter to Code Breakers by Craig P. Bauer (October 2, 2017); 5.1 Small Numbers Can Have Huge Impacts on Climate and Health by Naomi Oreskes (June 1, 2020); 5.2 The Heisenberg Uncertainty Principle of Social Science Modeling by Ben Klemens (July 7. 2020); 5.3 This Formula Calculates How Many Calories You Burn If You're Doing Absolutely Nothing by Florian Freistetter (July 27, 2021); 5.4 When Assessing Novel Risks, Facts Are Not Enough by Baruch Fischho (September 1, 2019); 6.1 Fractions: Where It All Goes Wrong by Robert S. Siegler (November 28, 2017); 6.2 Learning on the Back of an Envelope by Amanda Baker (October 10, 2018); 6.3 The Surprising Link between Recreational Math and Undecidability by Evelyn Lamb (December 7, 2019); 6.4 To Keep Students in STEM fields, Let's Weed Out the Weed-Out Math Classes by Pamela Burdman (March 15, 2022); 7.1 A Random Walk through the English Language by Jordan Ellenberg (June 11, 2021); 7.2 Farewell to the Fractional Foot by Evelyn Lamb (October 25, 2019); 7.3 For Math Fans: A Hitchhiker's Guide to the Number 42 by Jean-Paul Delahaye (September 21, 2020); 7.4 Forget Meters and Feet by Evelyn Lamb (January 23, 2020); 7.5 Math Is More Than Just Numbers: Celebrate Pi Day a Different Way by Alissa S. Crans (March 14, 2022); 7.6 Mathematicians Are Trying to 'Hear' Shapes by Rachel Crowell (June 28, 2022).

Each author biography was accurate at the time the article was originally published.

INDEX